QUANTUM TOPOLOGY
AND
GLOBAL ANOMALIES

ADVANCED SERIES IN MATHEMATICAL PHYSICS

Advanced Series in Mathematical Physics
Vol. 23

QUANTUM TOPOLOGY
AND
GLOBAL ANOMALIES

Randy A. Baadhio

Theoretical Physics Group, Physics Division
Lawrence Berkeley National Laboratory and Department of Physics
University of California, Berkeley

 **World Scientific**
Singapore • New Jersey • London • Hong Kong

Published by

World Scientific Publishing Co. Pte. Ltd.

P O Box 128, Farrer Road, Singapore 912805

USA office: Suite 1B, 1060 Main Street, River Edge, NJ 07661

UK office: 57 Shelton Street, Covent Garden, London WC2H 9HE

British Library Cataloguing-in-Publication Data
A catalogue record for this book is available from the British Library.

QUANTUM TOPOLOGY AND GLOBAL ANOMALIES

ISBN   981-02-2726-4
ISBN   981-02-2727-2 (pbk)

This book is printed on acid-free paper.

Printed in Singapore by Uto-Print

Preface

Few cross interactions in science have been as successful as that of physics and the mathematical branches of differential geometry, algebraic and low dimensional topology. This book is a contribution to the ongoing dialogue between these sciences, through the research of the author over the last seven years. The interaction between these subjects has been a dominant feature of research developments in the past few years, so much so in fact, that no single book can cover them all. I have chosen to present material on two specific instances in which such developments have occurred in the last decade or so and for which I have been, to some extent, involved. The material thus covered in this book centers on Chern-Simons-Witten theories, also known as topological quantum field theories, or, simply, as quantum topology. This topic constitutes the first part of the book, comprising eight chapters. They are written in a purely mathematical style and are intented to serve as an introduction to our physicist readership; we take the reader on a rapid trip through the elements of manifold theory, differential geometry, particularly the fundamentals of Lie groups, fibre bundles and connections, and algebraic topology with emphasis on mapping class groups; and homotopy and homology groups with emphasis on de Rham cohomology. Much of this material is brough together in chapters five, six, and seven.

Chapter one provides an introduction to three-manifold invariants with detailed analysis of the Chern-Simons invariant, the η-invariant. Mapping class groups are put to use in the description of three-manifolds invariant in chapter two using a construction by Khono which largely invokes the techniques of Moore and Segal on conformal field theories. Chapter three was written with the goal of providing introductory material to physicists on the relationship between Teichmüller space, moduli space, and mapping class groups. The interplay between them is discussed at great length, along

V

with the algebraic and homological characteristics they share. This material is extended to chapter four, where we investigate the cohomology of mapping class groups, and their p-torsion. We also study the question of how stable can mapping class groups be, using Harer's homological stability theorem. The full fledged role of mapping class groups in physics is discussed, of course, in chapter two, were we find a three-manifold invariant derived from it, but furthermore, in the second half of the book, namely in chapter 13 entitled Mapping Class Groups and Global Anomalies. The first part of this book ends with a chapter on the geometric quantization of Chern-Simons-Witten theories.

The second half of the book essentially deals with anomalies. We begin with an independent chapter on the relationship between deformation quantization and the occurence of global anomalies. The book then unfolds in an extensive study of chiral and gravitational anomalies, and anomalies and the index theorem, respectively in chapters 10 and 11. Global anomalies of both gauge and gravitational nature are introduced in chapter 12. Here, we investigate Witten's celebrated SU(2) global gauge anomaly, but also the manifestation and cancellation of global gravitational anomalies in some ten and six dimensional supergravity theories. A case study of global gravitational anomalies is made in the context of the ten dimensional heterotic string theory. Chapter 13 provides a detailed analysis of the manifestation of global anomalies in Chern-Simons-Witten theories in relation to three dimensional mapping class groups. The analysis of global gravitational anomalies in ten dimensional physics requires the existence of a special class of spheres, the so-called exotic spheres. Exotic spheres, namely the Milnor-Kervaire ones are explicitly used in Witten's formula for the anomaly cancellation as exhibited in chapters 12. Louis H. Kauffman has generously contributed, in its entirety, a chapter describing the construction of exotic spheres through characteristic classes. The interested reader is invited to read it in chapter 14.

Much is left unsaid in this book about the ongoing interaction between mathematics and physics. I would have loved to write on additional topics in which I have been involved, for instance about very exotic spheres and their role as gravitational instanton and/or soliton and the now available technique to detect them, or the role of classical invariant of knot–such as the Arf invariant– in detecting exotic structures, from dimensions three to eleven. But I realized that writing a monograph seems to invariably result in limiting one to a lesser scope.

The number of people that have contributed to this endeavor is truly astonishing. My gratitude goes to my friend and colleague Lou Kauffman for taking the time to write chapter 14; for reading the original manuscript and for providing critiques which have resulted in an improved manuscript, particularly chapters 8 and 9. I wish to acknowledge the expert proof-reading of Arthur Greenspoon of the American Mathematical Society. Arthur is to be thanked for several readings of the manuscript. The book, *Quantum Topology*, which I co-authored with Kauffman in 1993, benefited a great deal from Arthur's proof-reading work. His contribution is greatly appreciated here. Anne-Marie Piché is also to be thanked for her first draft corrections of early chapters writing. I had help from Zheng Huang, Gelato Sergio, and Orlando Alvarez in setting up the necessary OzTeX version of LaTeX and related software necessary for writing the entire manuscript on my Macintosh Powerbook. Their help is very much appreciated. I have been fortunate enough to find a stimulating and supportive environment at Berkeley. I therefore wish to thank Robert Cahn, my colleagues in the physics and mathematics departments, Irving Kaplansky and the staff of the Mathematical Sciences Research Institute for providing such a thriving environment.

The help of Gary Feldman, Michael Park, Steve O'Brien; U-Roy, Luanne Neumann, and H. O. Woba is greatly acknowledged as well. Flavio Robles and Gary Bergren are to be thanked for the illustrations.

I am grateful for support from the Eppley Foundation for Research Inc., the National Science Foundation, and the U.S. Department of Energy.

Berkeley, California, May 1996

Contents

Chapter 1

The Ongoing Quest For Three-Manifold Invariants

How much we can know about the universe in which we live depends very much on our ability to understand three-dimensional manifolds. This is a difficult task since three-manifolds are inherently complicated objects and there is a bewildering array of them. There are several methods currently available for constructing three-manifolds (e.g. the combinatorial approach, the Heegaard gluing approach, the Dehn surgery approach, etc.) What is lacking, however, are methods for determining when 3-manifolds are the same. Topologically speaking, two objects are said to be equivalent if there is a homeomorphism from one to the other. A homeomorphism is a one-to-one onto map that is continuous, and whose inverse is also continuous. In short, it is a topological isomorphism.

Over the last forty years, invariants have proven to be the most effective tools available for studying three-manifolds. The problem is the following: standard invariants, while useful, do not give us all of the information that we would like to have.

Nature, as we observed earlier, has blessed us with an abundant variety of 3-manifolds. According to Thurston's work [1], most 3-manifolds are hyperbolic. A hyperbolic 3-manifold is a three-manifold with a metric locally isometric to hyperbolic 3-space. A typical three-manifold is either topologically simple or otherwise possesses a hyperbolic structure. For instance, a knot in the 3-sphere possesses a hyperbolic complement , in which case it is referred to as a hyperbolic knot, when it is neither a torus knot nor a satellite knot. It should be noted that all but a finite number of the 3-manifolds obtained through performing Dehn surgery on a given hyperbolic knot have hyperbolic structures.

In essence, a 3-manifold must be topologically complicated to admit a hyperbolic structure. This may explain why the discovery of closed, hyperbolic 3-manifolds has been a slow process. Throughout this chapter and the book, when referring to a closed manifold, we shall mean a manifold that is compact and without boundary. Orientable shall mean that the manifold contains no mirror-reversing path.

The first cited construction of some hyperbolic 3-manifolds is credited to Löbell and dates as far back as 1931. Later, in 1933, Seifert and Weber discovered an easier way to describe hyperbolic 3-manifolds. But it would take about 38 years before Alan Best discovered, in 1971, several other hyperbolic 3-manifolds. Shortly thereafter, Bob Riley put hyperbolic structures on several knot and link complements, and Troels Jørgensen constructed hyperbolic three-manifolds that fiber over the circle. The Riley and Jørgensen examples were a major source of motivation for Thurston's seminal work on three-dimensional hyperbolic universes.

The evidence to date suggests that hyperbolic 3-manifolds are the most abundant, the most complicated, and the most important class of 3-manifolds. Therefore, for physicists and mathematicians alike, studying the 3-dimensional universe in which we live is best achieved by restricting our attention to topological three-manifolds which admit hyperbolic structures. Hyperbolic structures reveal many insights about 3-dimensional topology and physics. Among these is the discovery of a new class of 3-manifold invariants.

Below, we take up the quest for 3-manifold invariants. The latest The latest invariants to have surfaced in the past three years are presented in the next chapter.

Determining when two 3-manifolds are the same (that is, are homeomorphic) is a highly non-trivial procedure. In contrast, in dimension two, one can solve this problem by using a single invariant: the Euler characteristic. This invariant assigns a number to each closed, orientable 2-manifold. This, in turn, allows us to answer basic questions about such manifolds.

For instance, the problem of classifying 2-manifolds is made easier by the classical fact that two closed, orientable two-manifolds are homeomorphic if, and only if, they have the same Euler characteristic. Unfortunately, the Euler characteristic is not a viable invariant for 3-manifolds since all closed three-manifolds have Euler characteristic zero.

One may wonder at this point: are there are any other options? The answer is, yes. The fundamental group of a 3-manifold is one of those. It does contain far more information than the Euler characteristic, and is essentially straightforward to compute. It is, however, not considered to be a good invariant primarily because extracting information out of it is difficult. William Massey [2] provides us with a good example underlying this dilemma: given two group presentations (a presentation is a description of a group in terms of generators and relations), are they presentations of the same group or of different groups?

Homology groups, on the other hand, are often able to distinguish between 3-manifolds, and they are relatively easy to compute. As such, they are good candidates for invariants. But, there are several examples in which they fail to distinguish between 3-manifolds. All knot complements in the 3-sphere have the very same homology group, for instance.

Three of the most natural invariants of hyperbolic 3-manifolds are the volume, the Chern-Simons invariant, and the η-invariant.

1.1 Volume as a Three-Manifold Invariant

Consider a 3-manifold M^τ obtained via surgery from the manifold M. A theorem of Ruberman [3] states the following:

$$\text{vol}\,(M^\tau) \;=\; \text{vol}\,(M).$$

In essence, Ruberman's theorem implies that the volumes of hyperbolic 3-manifolds are qualitative invariants, since even a radical procedure such as surgery leaves them unchanged.

How does one compute the volume of a given hyperbolic 3-manifold? The starting point is perhaps to consider the analogous 2-dimensional case. In dimension two, the closest analogue to a hyperbolic structure is the Poincaré disk, a hyperbolic plane with infinitesimal metric

$$ds = \frac{\sqrt{(dx)^2 + (dy)^2}}{(1 - (x^2 + y^2))/2}.$$

An infinitesimal line segment parallel to the x-axis at the point (x, y) has hyperbolic length

$$\frac{dx}{(1 - (x^2 + y^2))/2};$$

similarly, a segment parallel to the y-axis has hyperbolic length

$$\frac{dy}{(1 - (x^2 + y^2))/2}.$$

Combining these, we see that an infinitesimal rectangle has hyperbolic area

$$\frac{dx\,dy}{(1 - (x^2 + y^2))^2/4}.$$

Thus, the hyperbolic area of a region in the Poincaré disk is computed by integrating the area form:

$$dA = \frac{dx\,dy}{(1 - (x^2 + y^2))^2/4}$$

over the region.

In dimension three, the Poincaré ball of hyperbolic 3-space has infinitesimal metric

$$ds = \frac{\sqrt{(dx)^2 + (dy)^2 + (dz)^2}}{(1 - (x^2 + y^2 + z^2))/2},$$

and volume form

$$dV = \frac{dx\,dy\,dz}{(1 - (x^2 + y^2 + z^2))^3/8}.$$

The volume of a tetrahedron can be computed by integration, and we obtain a formula for the volume of the tetrahedron in terms of its dihedral angles, as shown by Thurston [1]. As such, the problem of computing the volume of a hyperbolic 3-manifold reduces to the problem of decomposing it.

Is the volume a useful invariant for three-manifolds three-manifolds? It is in the sense that it is effective in distinguishing between 3-manifolds. Unfortunately, the volume is far from being a complete invariant for (hyperbolic) three-manifolds: there are examples of non-homeomorphic (i.e. distinct) hyperbolic 3-manifolds which nonetheless have equal volumes.

Provided that the volume fails, two remaining natural invariants for three-manifolds are the Chern-Simons invariant and the η-invariant. Both play a central role in various aspects of 3-dimensional physics, as we shall see in due course. Mostow's theorem is a sufficient criteria for proving that they are topological invariants for closed, hyperbolic 3-manifolds as well.

1.2   The Chern-Simons Invariant

In 1974, Chern and Simons [4] defined a certain 3-form Q on the oriented frame bundle $F(M)$. Owing to the fact that any orientable 3-manifold is parallelizable, one can focus on sections of the frame bundle, which act to pull Q back to M. We refer to this operation as a pull-back. Integrating s^*Q over M produces a real number; this number depends a priori on the choice of section of $F(M)$. Given one section, any other differs from it by a map from M to $SO(3)$. This means that the integral of the Chern-Simons form changes by $8\pi^2$, the degree of the map.

Thus, we can write the Chern-Simons invariant as:

$$CS\,(M) = \frac{1}{8\pi^2} \int_M s^*Q \ (mod\,1).$$

This invariant grew out of attempts by Chern and Simons to develop a combinatorial formula for the first Pontryagin number! of a given 4-manifold.

Though it has given us an enormous number of successful applications both in physics and mathematics, a drawback of the Chern-Simons invariant is that it is very difficult to compute. As a matter of fact, it was several years after its inception before we knew whether or not it was a trivial invariant for hyperbolic 3-manifolds. The first sign that the Chern-Simons invariant was not trivial for this class of manifolds surfaced in 1981. But it wasn't until

1986 that any strong evidence emerged corroborating its non-triviality; this was done by Meyerhoff in reference [5]. His approach can be summarized as follows: in the circle $\mathbb{R}/\mathbb{Z}$, the Chern-Simons invariant takes on a dense set of values. The trick is then to investigate the geometrical behavior of the set in question. We refer the interested reader to [5] for more details.

There exists an analytic relation between the volume invariant and the Chern-Simons invariant for hyperbolic 3-manifolds. This property was discovered in 1986 by Walter Neumann and Don Zagier [6].

Just how useful is the Chern-Simons 3-form as a three-manifold invariant? Or, put differently, can the Chern-Simons invariant enable one to distinguish large classes of hyperbolic 3-manifolds that have equal volumes? In answer, Meyerhoff and Ruberman [7] offer the following:

Theorem 1.1 (Meyerhoff-Ruberman) *Consider the circle $\mathbb{R}/\mathbb{Z}$ in which the Chern-Simons invariant takes its value. Given any rational number in $\mathbb{R}/\mathbb{Z}$, there exist hyperbolic! three-manifolds with equal volumes whose Chern-Simons invariants differ by this rational number.*

This is undoubtedly a nice result: one can appreciate the degree of refinement for which a difference in values between the two invariants reduces to a computable, yet easily quantifiable and manageable norm. The problem, however, is simply that a systematic understanding of manifolds with equal volumes and different Chern-Simons invariants seems a long way off.

1.3   The η-invariant

Most of the invariants we just have just encountered have, to some degree, some drawbacks. As such, they are in no way the ultimate sought 3-manifold invariants. Another possibilty is to look at the η-invariant. This invariant was introduced by Atiyah, Patodi and Singer [8]. In its original formulation for odd-dimensional manifolds, it was given in terms of the eigenvalues of the Laplace operator. It was, furthermore, initially introduced to measure the extent to which the Hirzebruch signature formula fails for geometric 4-manifolds with boundary.

Atiyah, Patodi and Singer gave the following remarkable formula [8], which we take as the definition of the η-invariant.

Theorem 1.2 (Atiyah-Patodi-Singer) *Consider a 4-dimensional manifold W whose boundary is a 3-dimensional manifold M. Choose a framing α on M, such that it gives rise to the Pontrjagin number $p_1(W)$. Define the signature defect $\sigma(M, \alpha)$ to be the integer $\frac{1}{3} p_1(W) - \text{sign}(W)$. We can then write the η-invariant as*

$$\eta(M) = \frac{1}{12\pi^2} \int \alpha^\star Q + \sigma(M, \alpha). \qquad (1.1)$$

The η-invariant, it should be noted, is closely related to the Chern-Simons invariant; specifically,

$$3\eta(M) = 2CS(M) \bmod \mathbb{Z}.$$

The η-invariant contains information that the Chern-Simons invariant does not have. For instance, there are examples of hyperbolic 3-manifolds with Chern-Simons invariants equal but different η-invariants. Tomoyoshi Yoshida has made substantial progress toward a systematic computation of the η-invariant for hyperbolic 3-manifolds [9].

1.4 The Chern-Simons Invariant Revisited

The discovery by Vaughan Jones of a new polynomial invariant of links in the 3-sphere in 1985 [13] was an important breakthrough which has led to the introduction of a whole range of new techniques in three-dimensional topology. The original Jones polynomial, a Jones polynomial in one variable, was obtained via a braid description of a link, utilizing the remarkable properties of some representations of the braid group which arose in the theory of von Neumann algebras. Early developments were largely combinatorial, leading to alternative definitions of the invariant and to generalizations, including a two-variable polynomial which specializes in both the Jones polynomial and the classical Alexander polynomial after appropriate substitutions.

The new invariants are comparatively easy to calculate and have had many concrete applications, but for some time no satisfactory conceptual definition of the invariants was known-one not relying on the special combinatorial presentations of a link. It was not clear, for example, whether such invariants could be defined for links in other 3-manifolds. While there were many intriguing connections between the Jones theory and statistical mechanics, for instance through the Yang-Baxter equation and the newly developed theory of quantum groups, it was a major problem to find the correct geometric setting for the Jones theory.

In July of 1988, Witten proposed a scheme which largely resolved this problem. He showed that the invariants (including the Kauffman polynomial) should be obtained from a quantum field theory with a Lagrangian involving the Chern-Simons invariant of connections. Witten's approach [10] provided a truly natural definition of the invariants, and indeed allowed considerable generalization to links in arbitrary three-manifolds.

Taking in particular the empty link, he obtained a new invariant of closed 3-manifolds. The challenge in this approach arose from the notorious difficulties of quantum field theory in attaching a real meaning to the functional integral over the space of connections which is involved.

The principal source of interest in this new class of theories was the realization that despite these foundational difficulties, a new class of invariants could be constructed and yet make concrete predictions which could be verified on a more elementary level.

The definition of these invariants is given in terms of a functional integral, namely

$$Z_{k,G}(M,g) = \int exp\,(ik\,\mathrm{CS}\,(A))\,\mathcal{D}[A], \qquad (1.2)$$

where CS denotes the Chern-Simons functional, and G is a compact semisimple, simply connected Lie group, g is a metric on M; the integration is done over all gauge equivalence classes of connections on a principal G-bundle over the 3-manifold M.

The functional integral (1.2) is, however, not a mathematically well defined object, and this very fact underscores the discovery of a new class of 3-manifold invariants–some of which are presented in the next chapter.

How do physicists avoid this problem? Mainly by using the following two standard methods. The first method consists of giving a proper definition of functional integrals via perturbation theory. With regard to equation (1.2), this would be the limit $k \rightarrow \infty$. In this limit, one can try to compute the integral by the stationary phase approximation method. Witten, in reference [10], gives the following formula for the large k limit of (1.2):

$$Z_{k,G}(M,g) = \exp i\pi \dim G \left(\frac{\eta_{\mathrm{grav}}}{2} + \frac{1}{12} \frac{\mathrm{CS}(A^g)}{2\pi} \right)$$
$$\Sigma_{[A^0]} \exp i \left(k + \frac{c_2(G)}{2} \right) \mathrm{CS}(A^0) T_{A^0};$$

(1.3)

where the sum is taken over the gauge equivalence classes of flat connections on the principal G-bundle over M, A^g is the corresponding Levi-Civita connection, η_{grav} is the η-invariant of the operator $\star D^g + D^g \star$ (D^g being the exterior derivative twisted by A^g), c_2 is the value of the quadratic Casimir operator in the adjoint representation of G, and finally, T_{A^0} is the Ray-Singer torsion of A^0.

The sum in (1.3) is defined if $\{[A^{(0)}]\}$ is a set of isolated points; otherwise the sum should be replaced by an integral over the classes of flat connections, with a suitable measure. A coherent description of the asymptotics of (1.2) in this case is an interesting and, as far as we know, open problem. It has been shown by Witten that the right hand side of (1.2) should depend only on the 2-framing of the 3-manifold M. On the other hand, Atiyah's canonical framing [11] should yield a similar Atiyah's canonical framing 3-manifold invariant.

The second way to rigorously define the functional integral (1.2) is to use some phenomenological formula for studying (or rather to find) transformation properties of (2) under certain natural transformations. In most cases, these properties uniquely fix the object on the left hand side of (1.2), and give an independent rigorous definition of it. Reshetikhin and Turaev's definition of 3-manifold invariants via surgery on a link in S^3 [12] is perhaps the most readily available realization of this program for the functional integral (1.2).

Equation (1.3) tells us that in order to find the limit of (1.2) for $k \rightarrow \infty$, we have, at least, to sum over all classes of flat connections. On the other hand, it is clear that each individual term in (1.2) should be an invariant of the pair $\left(M, [A^{(0)}] \right)$. After appropriate normalization, it becomes an element of $\mathbb{C}\left[\left[\frac{1}{k} \right] \right]$, which we write as $Z_k \left(M, [A^{(0)}] \right)$. Axelrod and Singer [14], and Kontsevich [15] found that this power series exists if M is a rational homology sphere. There are additional factors to be taken into account, namely, the de Rham complex twisted by $A^{(0)}$ is acyclic (in Kontsevich's paper [15], $A^{(0)} = 0$; furthermore, the coefficients are a linear combination of integrals $\int_{M \times \cdots \times M} \omega$ for some suitable forms ω. These are invariants of pairs $\left(M, [A^{(0)}]\right)$. The term $Z_k (M, 0) \in \mathbb{C}\left[\left[\frac{1}{k}\right]\right]$ turns out to be an invariant of 3-manifolds.

This term differs from the 3-manifold invariants obtained via canonical framing, and for $k \in \mathbb{N}$ seems to exist for rational homology spheres. Reconciling these two invariants is at the core of understanding perturbative Chern-Simons-Witten theories.

1.5 Outlook and Summary

Let us review the dilemma with which we are faced. We live in a three-dimensional universe of which very little is known in terms of its mathematical and physical properties. Although this universe is allowed to have infinitely many shapes, the message from Thurston is clear: hyperbolic 3-universes are the most abundant, important and yet complicated collection of all types of 3-manifolds.

In order to understand the world we live in, we need some objects whose primary role is to take the unwieldy collection of information that defines the universe, and distill it into a manageable packet. Such objects are called invariants.

For hyperbolic 3-manifolds, we have seen that neither the homology nor the fundamental group are satisfactory invariants. New invariants are crucially needed. If we focus our attention on 3-dimensional hyperbolic universes with finite volume, then a fundamental theorem of Mostow tells us that such invariants ought to be topological invariants.

Hyperbolic 3-manifolds have natural invariants: the volume, the Chern-Simons invariant, and the η-invariant. The volume, we have seen, has proven successful at distinguishing between manifolds with the same homology, while the Chern-Simons invariant and the η-invariant can distinguish between many hyperbolic 3-universes with the same volume. In addition, these invariants should be able to yield information about the underlying universe. For instance, the volume appears to be a good measure of complexity, while the Chern-Simons invariant appears to measure handedness.

A fundamental issue is to measure to what extent the volume, the Chern-Simons and the η-invariant, when taken together, determine a given 3-dimensional universe. Unfortunately, as is often the case with central questions there is as yet no known answer.

To further complicate our quest to understand the universe in which we live, some examples have recently surfaced of 3-manifolds that are not distinguishable by the three invariants we just mentioned. Cusped hyperbolic 3-manifolds fall into this category. These are hyperbolic mutant manifolds [7]; they do share among themselves equal volumes and Chern-Simons invariant. But the nature of the η-invariant is problematic, that is to say, hard to compute. A mutation is the result of cutting a 3-manifold M along a genus two surface Σ_g, and regluing via the (unique) involution in the center of Σ_g.

It is a likely possibility that mutant manifolds are insensitive to the above-mentioned three invariants, in part because of our poor knowledge of 3-dimensional mapping class groups. The unique involution defining the pasting of the mutant manifold originates from the center of the mapping class group. What is not known at this point, however, is the nature of this involution in terms of the mapping class group itself. Depending on what subgroup of the mapping class group determines the mutation, we may have an impetus to investigate 3-dimensional mapping class groups and subgroups more aggressively.

Mapping class groups enhance our interest in quantum topology because of their central roles in various forms of quantization, operator ordering, global anomalies [16], etc. A great many of these issues will be discussed throughout this book. The next chapter is devoted to an invariant of 3-manifolds obtained using mapping class groups, while the relation between mapping class groups, Teichmüller space and moduli space is thoroughly presented in following chapters.

We should point out that mutant hyperbolic three-manifolds are not insensitive only to the η-invariant: the volume, the Chern-Simons and η-invariant do not provide a complete set of invariants for closed, hyperbolic, mutant 3-manifolds. Certain mutations, such as the one described by Meyerhoff and Ruberman in reference [7], leave the volume, the Chern-Simons invariant (mod 1), and the η-invariant unchanged.

Perhaps it comes across to the reader that we critically need new invariants of three-manifolds. If so, then the objective of this chapter to convey the efforts behind the continuous quest for three-manifold invariants has been achieved.

1.6   References

[1] Thurston, William: *The Geometry and Topology of 3-Manifolds*, Princeton University Press (1978).

—*Three-Dimensional Manifolds, Kleinian groups and hyperbolic geometry*, Bull. American Mathematical Society (2) 6 (1982) 357-381.

[2] Massey, Williams: **Algebraic Topology: An Introduction**, Harcourt, Brace, and World Inc., (1967) New York, p. 106.

[3] Ruberman, Daniel: *Mutations and Volumes of Knots in S^3*, Inventiones Math. 90 (1987), 189-216.

[4] Chern, S. S. and Simons, J.: *Characteristic Forms and Geometric Invariants*, Annals Math. 2, 99 (1974) 48-69.

[5] Meyerhoff, Robert: *Density of the Chern-Simons Invariant for Hyperbolic 3-manifolds,*in **Low Dimensional Topology and Kleinian Groups**, Editor: D. B. A. Epstein, Cambridge University Press, (1986) 217-239.

[6] Neumann, Walter and Zagier, Don: *Volumes of Hyperbolic Three-manifolds*, Topology 24 (1985) 307-332.

[7] Meyerhoff, Robert and Ruberman, Daniel: *Mutation and the η-invariant*, Journal of Differential Geometry 31 (1990) 101-130.

[8] Atiyah, M. F., Patodi, V. K. and Singer, I.: *Spectral Asymmetry and Riemannian Geometry* I: Math. Proc. Camb. Phil. Soc. 78 (1975) 405-432.

[9] Yoshida, Tomoyoshi: *The η-invariant of Hyperbolic 3-Manifolds*, Invent. Math. 81 (1985) 473-514.

[10] Witten, Edward: *Quantum Field Theory and the Jones Polynomial*, Comm. Mathematical Physics 121 (1989) 351-399.

[11] Atiyah, M. F.: *On Framings of 3-Manifolds*, Topology 29 (1990) 1-7.

[12] Reshetikhin, N. Y. and Turaev, V. G.: *Invariants of 3-Manifolds via Link Polynomials and Quantum Groups*, Invent. Math. 103 (1991) 547-597.

[13] Jones, Vaughan: *A New Polynomial Invariant for Links via von Neumann Algebras*, Bull. Amer. Math. Soc. 129 (1985) 103-112.

[14] Axelrod, S. and Singer, I.: *Perturbation Theory for Chern-Simons Theory*, in **Proc. *XX* Int. Conf. on Diff. Geom. Methods in Physics**, World Scientific (1991) 3-45.

[15] Kontsevich, M.: *Feynman Diagrams and Low-dimensional Topology*, prepint, 1992. To appear in the Proceedings of the First European Congress of Mathematicians.

[16] Baadhio, R. A.: *Mapping Class Groups for D = 2+1 Quantum Gravity and Topological Quantum Field Theories*, Nuclear Physics B441 (1995) 383-401.

Chapter 2

Mapping Class Groups and 3-Manifold Invariants

The following exposition is aimed at exhibiting one of the main roles played by the mapping class group in quantum topology. We shall describe a new type of 3-manifold invariant derived from the mapping class group. This invariant surfaced in late 1990; it was discovered by Kohno [1] and requires quite a bit of interplay between the Knizhnik-Zamolodchikov monodromy equation as well as the Moore-Seiberg conformal field theoretic approach to fusing and braiding matrices.

The procedure also makes use of the Heegaard decomposition of 3-manifolds; the trick here being to show that the holonomy induced by the Knizhnik-Zamolodchikov equation gives rise to solutions of polynomial equations. An appropiate set of rules is then applied to obtain a projective linear representation of the mapping class group, that is, of the group of isotopy classes of orientation preserving self-diffeomorphisms. We write this representation as

$$\rho_K : \pi_0 \, \mathrm{Diff}^+ \left(M_g \right) \to \mathrm{GL} \left(Z_K \left(\gamma \right) \right) / \Gamma_K;$$

where Γ_K is the cyclic group generated by $2\pi \, i \frac{c}{24}$, with $c = \frac{3K}{K+2}$, and $\pi_0 \, \mathrm{Diff}^+ \left(\Sigma_g \right)$ denotes the mapping class group of the Riemann surface Σ_g of genus g; moreover, $Z_K \left(\gamma \right)$ is a finite dimensional complex vector space and γ denotes the dual graph of pants decomposition of Σ_g. The finite

14

dimensional vector space $Z_K(\gamma)$, it turns out, appears in a combinatorial description of the space of conformal blocks for an SU (2) Wess-Zumino-Witten model at level K. This can be shown using Tsuchiya, Ueno and Yamada's work outlined in reference [2].

In order for us to describe this new invariant, we first need to explain in great detail where the connections with conformal field theory lie. Additional topological properties will come out of this.

2.1   Markings on Riemann Surfaces

Consider a closed, orientable two-dimensional surface Σ_g of genus g; define in Σ_g a maximal collection of disjoint non-contractible and pairwise non-isotopic smooth circles. do Carmo provides in his book [3] beautiful examples of such constructions. The isotopy classes of these circles are referred to as a marking of Σ_g. When $g > 1$, a classification theorem in Riemannian geometry tells us that there are exactly $3g - 3$ circles whose complementary space consists of $2g - 2$ trinions (these are 3-holed spheres).

We now move to complicate the process a bit. To a marking μ of Σ_g, we associate a dual graph $\gamma(\mu)$. According to results of Hatcher and Thurston [4], any two markings μ_1 and μ_2 of Σ_g can be obtained from one another by an appropriate use of the mapping class group. Let us call this procedure a move. There exist several types of moves, but the ones of interest to us are the moves that induce a change in the associated graph (that is, moves for which the mapping class group has a non-trivial action.)

Moves that induce a change in the associated graph are called elementary fusing operators. Following Kohno, we pick a (positive) integer K, which we refer to as the level. To the dual graph $\gamma(\mu)$, we further associate a finite dimensional complex vector space. In order for us to define the latter, let us pick a given set of half-integers, labelled $\mathcal{P}_K$:

$$\mathcal{P}_K = \left\{0, \frac{1}{2}, 1, \cdots, \frac{K}{2}\right\}.$$

Given a function

$$f : \text{edge}\,(\gamma(\mu)) \rightarrow \mathcal{P}_K,$$

we say that it is an admissible weight of the graph $\gamma(\mu)$ if it can be shown to satisfy Clebsch-Gordan condition for SL $(2, \mathbb{C})$ [5]:

$$|f(c_1) - f(c_2)| \leq f(c_3) \leq f(c_1) + f(c_2),$$

$$f(c_1) + f(c_2) + f(c_3) \in \mathbb{Z}.$$

To resume what we have done so far: take two irreducible representations of SL $(2, \mathbb{C})$ and project them into some irreducible components. The matrix elements of the projection are the Clebsch-Gordan coefficient. We are now in a position to define the finite dimensional complex vector space. This is the space $Z_K(\gamma(\mu))$ whose basis consists of the set of admissible weights of $\gamma(\mu)$. We write the basis of $Z_K(\gamma(\mu))$ as $\left\{e_{\gamma(\mu),f}\right\}_f$.

The properties of the marking can be extended to an orientable Riemann surface Σ_g with n holes $a_1, \cdots, a_n$ as follows. Observe that a marking μ consists of $3g - 3 + n$ circles. Fixing half-integers $j_1, \cdots, j_n \in \mathcal{P}_k$, we take $Z_K(\gamma(\mu); a_1, \cdots a_n; j_1, \cdots, j_n)$ to be a complex vector space whose basis is now labelled by the set of admissible weights f satisfying the consistency condition

$$f(a_k) = j_k, \quad 1 \leq k \leq n.$$

The generalization of marking to n holes reveals, in the process, a rather attractive physical connection: the vector space $Z_K(\gamma(\mu))$ corresponds to a combinatorial description of the conformal blocks of the SU (2) Wess-Zumino-Witten model at level K. An elegant description of this relationship is given in [6].

2.2   Using the Knizhnik-Zamolodchikov Equation

The construction of the 3-manifold invariant $\Phi_K(M)$ requires several ingredients. Among these is the Knizhnik-Zamolodchikov equation. The purpose of this section is to explain, along the lines of reference [7], how the basic properties of braiding and fusing matrices can arise from the holonomy of the Knizhnik-Zamolodchikov equation.

Several years ago, Moore and Seiberg, motivated by consistency conditions in conformal field theory, derived a series of polynomial equations among fusing matrices, braiding matrices and switching operators. The task before us is therefore to show that the holonomy of the Knizhnik-Zamolodchikov equation really provides a solution to the Moore-Seiberg polynomial equations.

Let us begin by choosing a representation of $SL(2, \mathbb{C})$, which is given by the half-integer spin j; call this representation V_j; it is an irreducible representation of dimension $2j + 1$. Note that $j_k \in \mathcal{P}_K$, that is, j_k is a part of the admissible weight. We now pick an orthonormal basis $\{I_\mu\}$ of $SL(2, \mathbb{C})$ with respect to the Cartan-Killing form. This gives:

$$\Omega = \sum_\mu I_\mu \otimes I_\mu.$$

The matrices Ω_{ij}, $1 \leq i, j \leq n$ are defined by

$$\Omega_{ij} = \sum_\mu \pi_i(I_\mu)\, \pi_j(I_\mu) \in \text{End}\,(V_{j_1} \otimes \cdots \otimes V_{j_n});$$

where π_i and π_j stand for the operation on the ith and jth components of the tensor product, respectively.

We are now in position to define the Knizhnik-Zamolodchikov equation. It is given by the formula

$$\frac{\partial \Phi}{\partial z_i} = \frac{1}{K+2} \sum_{j \neq 1} \frac{\Omega_{ij}}{z_i - z_j}\, \Phi, \ 1 \leq i \leq n, \qquad (2.1)$$

for a function $\Phi(z_1, \cdots, z_n)$ given over the space $X_n = \{(z_1, \cdots, z_n) \in \mathbb{C}\}$.

Consider a vector bundle over X_n with fiber $V_{j_1} \otimes \cdots \otimes V_{j_n}$. A solution of the Knizhnik-Zamolodchikov equation corresponds to a horizontal section of the integrable connection

$$\omega = \frac{1}{K+2} \sum_{1 \leq i < j \leq n} \Omega_{ij}\, d\log(z_i - z_j). \qquad (2.2)$$

The tensor product, $V_{j_1} \otimes V_{j_2}$ can be decomposed into a direct sum

$$V_{j_1} \otimes V_{j_2} = \bigotimes_j V_j$$

as SL $(2, \mathbb{C})$-modules. This sum ranges over all j satisfying the Clebsch-Gordan rule, i.e. $|j_1 - j_2| \le j \le j_1 + j_2$, where $j_1 + j_2 + j \in \mathbb{Z}$. The connection ω acts in a natural way on

$$\text{Hom}_{\text{SL}(2,\mathbb{C})} \left(V_{j_1} \otimes \cdots \otimes V_{j_n}, V_{j_{n+1}} \right). \tag{2.3}$$

2.2.1   The Fusing Matrices

Because the Knizhnik-Zamolodchikov equation possesses an analytic continuation, one can write the matrix

$$F \begin{pmatrix} j_2 & j_3 \\ j_1 & j_4 \end{pmatrix} = \left(F_{ij} \begin{pmatrix} j_2 & j_3 \\ j_1 & j_4 \end{pmatrix} \right)_{ij}. \tag{2.4}$$

It is defined in such a way that the solutions are related by

$$\Phi_{\gamma_1, j} = \sum_i F_{ij} \begin{pmatrix} j_2 & j_3 \\ j_1 & j_4 \end{pmatrix} \Phi_{\gamma_2, i}. \tag{2.5}$$

The above matrix is called the fusing matrix.

2.2.2   The Braiding matrices

A normalized (with respect to Wigner's $3j$-symbols [9]) solution of the Knizhnik-Zamolodchikov equation corresponds to the (half) monodromy of the equation with values in (2.3). The braiding matrix is then given by

$$\sigma^\star \Phi_{\gamma_1, j} (z_1, z_2, z_3) = \sum_i B_{ij} \begin{pmatrix} j_2 & j_3 \\ j_1 & j_4 \end{pmatrix} \Phi_{\gamma_1, i} (z_1, z_3, z_2). \tag{2.6}$$

Note that $\sigma^\star \Phi_{\gamma_1, j} (z_1, z_2, z_3)$ is an analytic continuation with respect to the elementary braid switching σ_i.

2.2.3   The Switching Operators

In addition to the braiding and fusing matrices, we consider the switching operator $S_K (j)$. This operator is to be constructed in such a way that it depends on K and j only, and furthermore, that it represents the action of the mapping class group $\pi_0 \, \text{Diff}^+ \, (\Sigma_g)$ on the vector space $Z_K \, (\gamma(\mu))$. Using techniques developed by Moore and Seiberg, we choose $[n]$, a q-integer:

$$[n] = \frac{q^{n/2} - q^{-n/2}}{q^{1/2} - q^{-1/2}},$$

with $q = \exp\left(2\pi \sqrt{-1}/(K+2)\right)$. Rewriting (2.4) gives

$$F_{00} \begin{pmatrix} k & k \\ k & k \end{pmatrix} = \frac{(-1)^{2k}}{[2k+1]}.$$

This allows us to derive the switching operator:

$$S_K \, (0)_{ij} = \sqrt{\frac{2}{K+2}} \, \sin \frac{(2i+1)(2j+1)}{K+2}, \quad 0 \leq i,j \leq K/2. \qquad (2.7)$$

According to Kac and Peterson's work [9], (2.7) represents nothing less than the modular properties of the characters, $\chi_j \, (\tau)$, of the integrable highest weight modules of level K, of the Lie algebra, of type $A_1^{(1)}$. This translates more appropiately into the relations

$$\chi_j \left(-\frac{1}{\tau}\right) = \sum_i S_k \, (0)_{ij} \, \chi_i \, (\tau);$$

$$\chi_j \, (\tau + 1) = \exp 2\pi \sqrt{-1} \left(\Delta_j - \frac{c}{24}\right) \chi_j \, (\tau).$$

Note that Δ_j stands for $\frac{j(j+1)}{K+2}$, and $c = \frac{3K}{(K+2)}$. For small values of K, in particular for $K = 1$, $S_K \, (j) \, (j = 0,1,2)$ takes on the simple form

$$S_1 \, (0) = \frac{1}{\sqrt{2}} \begin{pmatrix} 1 & 1 \\ 1 & -1 \end{pmatrix}.$$

And for $K = 2, 3$, we have:

$$S_2 \, (1) = \exp \left(-\frac{3\pi \sqrt{-1}}{4}\right);$$

$$S_3 \, (1) = -\frac{1}{\sqrt{2}} \exp \frac{3\pi \sqrt{-1}}{10} \begin{pmatrix} 1 & 1 \\ 1 & -1 \end{pmatrix}.$$

The following generalized formula for the switching operator is due to Li and Yu [10]:

$$S(j)_{pq} = \sum_k \exp 2\pi \sqrt{-1} \left(\Delta_k - \Delta_p - \Delta_q\right) S(0)_{0k} B_{pq} \begin{pmatrix} i & k \\ q & p \end{pmatrix}. \quad (2.8)$$

We pause to point out that solutions for general K can also be obtained independently from the SU(2) Wess-Zumino-Witten model over a 2-holed torus at level K.

2.3   Mapping Class Group Representations

Using the above defined fusing and braiding matrices and switching operators, we are now in position to construct the last remaining ingredients necessary to construct Kohno's 3-manifold invariant, i.e. a projective linear representation of the mapping class group. The starting point is the critical value $c = \frac{3K}{K+2}$. In this case,

$$\exp 2\pi \sqrt{-1} \left(\frac{c}{24}\right) \cdot \mathrm{id} \in \mathrm{GL}\left(Z_K(\gamma)\right)$$

generates a cyclic group, call it Γ_K, which in turn yields the following well-defined homomorphism:

$$\rho_K : \pi_0 \mathrm{Diff}^+ (\Sigma_g) \to \mathrm{GL}\left(Z_K(\gamma)\right)/\Gamma_K. \quad (2.9)$$

In order to completely determine the invariant $\Phi_k(M)$, some additional steps are required. Let $h \in \pi_0 \mathrm{Diff}^+ (\Sigma_g)$. This gives an equivalence relation

$$\rho_K(h) e_{\gamma,0} = \rho_K(h)_{00} e_{\gamma,0} + \sum_{f \neq 0} \rho_K(h)_{f,0} e_{\gamma,f},$$

where $e_{\gamma,0}$ is an element of the basis $Z_K(\gamma)$ corresponding to the admissible weight

$$f : \mathrm{edge}(\gamma) \to \mathcal{P}_K.$$

Put differently, one may think of $\rho_K(h)_{00}$ as the $(0,0)$-entry of the matrix $\rho_K(h)$ with respect to the basis $\{e_{\gamma,f}\}_f$.

The composition of fusing matrices induces an isomorphism of the vector spaces

$$Z_K\left(\gamma\left(\mu_1\right)\right) \cong Z_K\left(\gamma\left(\mu_2\right)\right).$$

Recall that $\mu_{1,2}$ are markings of Σ_g. The relation we have just written has an interesting consequence: it implies that the linear representation $\rho_K(h)_{00}$ does not depend on the choice of a marking.

2.4   The 3-Manifold Invariant $\Phi_K(M)$

We are now poised to define the topological! invariant $\Phi_K(M)$. The 3-manifold M is taken to be closed, orientable, and admitting a Heegaard decomposition. By appropriately using the projective linear representation ρ_K of the mapping class group, we find that this invariant is explicitly given by the formula

$$\Phi_K(M) = \left(\sqrt{\frac{2}{K+2}}\sin\frac{\pi}{K+2}\right)^{-g}\rho_K(h)_{00}. \tag{2.10}$$

This is a proper topological invariant with a fairly wide spectrum of universality. For instance, two closed, orientable 3-manifolds, M_1 and M_2, have similar invariants (i.e. $\Phi_K(M_1) = \Phi_K(M_2)$) provided that there is an orientation preserving homeomorphism between M_1 and M_2. We recall that a homeomorphism is a topological isomorphism.

Kohno's 3-manifold invariant seems to work well for manifolds made up of connected sums. In essence, he proved that

$$\Phi_K(M_1 \natural M_2) = \Phi_K(M_1) \cdot \Phi_K(M_2) \bmod \Gamma_K. \tag{2.11}$$

2.5   References

[1] Kohno, T.: *Topological Invariants for 3-Manifolds Using Representation of Mapping Class Groups I*, Topology, Vol. 31 No. 2 (1992) 203-230.

[2] Tsuchiya, A., Ueno, K., and Yamada, Y.: *Conformal Field Theory on Universal Families of Stable Curves with Gauge Symmetries*, Advanced Studies in Pure Math. Vol. 19 (1989) 459-566.

[3] do Carmo, M.: **Riemannian Geometry**, Birkhaüser, Boston, 1992.

[4] Hatcher, A., and Thurston, W.: *A presentation for the Mapping Class Group of an Orientable Surface*, Topology Vol. 19 (1980) 221-237.

[5] Alvarez-Gaumé , L., Sierra, G., and Gomez, C.: *Topics in Conformal Field Theories*, in **Physics and Mathematics of Strings**, World Scientific, 1990. New Jersey, Singapore, London.

[6] Knizhnik, V. G., and Zamolodchikov, A. B.: *Current Algebra and Wess-Zumino Models in Two Dimensions*, Nuclear Physics B247 (1984) 83-103.

[7] Moore, G., and Seiberg, N.: *Classical and Quantum Conformal Field Theory*, Commun. Math. Physics 123 (1989) 177-254.

[8] Kirillov, A. N., and Reshetikhin, N. Y.: *Representations of the Algebra* U_g (SL (2, $\mathbb{C}$)), *q-orthogonal Polynomials and Invariants of Links*, in **Infinite Dimensional Lie Algebra and Groups**, (1988) 285-342, World Scientific. New Jersey, Singapore, London.

[9] Kac, V. G. and Peterson, D. H.: *Infinite Dimensional Lie Algebras, Theta Functions and Modular Forms*, Advances in Math. 53 (1984) 125-264.

[10] Li, M. and Yu, M.: *Braiding Matrices, Modular Transformations and Topological Field Theories in* 2 + 1 *Dimensions*, Commun. Math. Physics 127 (1990) 195-224.

Chapter 3

Teichmüller Spaces and Mapping Class Groups

Teichmüller space, moduli space and mapping class groups are central objects in quantum topology. Since we shall devote several chapters to issues in which they play a central role, it is thus essential that we start with a solid mathematical background. The following exposition and those to come serve as this foundation.

Moduli space can be approached in several distinct ways. In particular, when the genus g is greater than one, it is the space of isometry classes of hyperbolic metrics on a surface; it is also the space of conformal equivalence classes of Riemann surfaces, and the space of algebraic curves of genus g, up to isomorphism.

These definitions reveal an elaborate interplay between hyperbolic geometry, complex analysis, algebraic topology and physics. As for the mapping class group, it acts properly discontinuously on the Teichmüller space with quotient the moduli space. Hence the rational cohomology of the moduli space may be identified with that of the mapping class group. Not only does this add a topological and algebraic perspective to things, but furthermore, it elevates the physical interest in understanding the interplay between such central objects to higher ground!

3.0.1 The Moduli Space

Let Σ_g be a closed, oriented surface of genus g. The primary object of interest to us is the space which parametrizes all the conformal structures carried by the surface Σ_g. This space is referred to as the moduli space of the surface Σ_g; we shall denote it by $\mathcal{M}_g$. The moduli space $\mathcal{M}_g$ can also be defined as the space of hyperbolic metrics on Σ_g, or, as the space of algebraic curves of genus g, up to appropriate notions of equivalence.

3.0.2 The Teichmüller Space

Here, we take a conformal point of view. Define a marked Riemann surface to be the pair $(\Sigma_g, [f])$. Σ_g is defined as above and $f : \Sigma_g \to \Sigma_g$ is a homeomorphism; $[f]$ stands for the homotopy class of f. Two marked Riemann surfaces $\left(\Sigma_g^1, [f_1]\right)$ and $\left(\Sigma_g^2, [f_2]\right)$ are said to be equivalent if there is a conformal homeomorphism $h : \Sigma_g^1 \to \Sigma_g^2$ such that $[f_2 \circ h] = [f_1]$. The collection of equivalence classes is denoted by $\mathcal{T}_g$; it has a natural topology (which will soon be made explicit), and is called the Teichmüller space of genus g [1, 2].

3.0.3 Moduli Space and the Mapping Class Group

From a Teichmüller space perspective, the moduli space $\mathcal{M}_g$ can be obtained in a rather different fashion, that is, by omitting the marking $[f]$. To make this point a bit more detailed, we introduce the mapping class group, $\pi_0 \operatorname{Diff}^+ (\Sigma_g)$, which we abbreviate as Γ_g. This group consists of homotopy classes (or equivalently, isotopy classes) of orientation preserving homeomorphisms (resp. self-diffeomorphisms) of Σ_g. The formula,

$$[g] \cdot (\Sigma_g, [f]) = (\Sigma_g, [gf]), \qquad (3.1)$$

defines an action of Γ_g on $\mathcal{T}_g$; the resulting moduli space of conformal structures on Σ_g is denoted once again by $\mathcal{M}_g$. The second method to describe $\mathcal{T}_g$ and $\mathcal{M}_g$ is to use hyperbolic geometry. This approach makes sense only when $g > 1$. By a hyperbolic surface, we shall mean a smooth surface Σ_g, equipped with a complete Riemannian metric of constant curvature -1. A marked hyperbolic surface is given by the pair $(\Sigma_g, [f])$, where $f : \Sigma_g \to \Sigma_g$

is a homeomorphism. We say that $\left(\Sigma_g^1, [f_1]\right)$ is equivalent to $\left(\Sigma_g^2, [f_2]\right)$ if there is an isometry $h : \Sigma_g^1 \to \Sigma_g^2$ with $[f_2 \circ h] = [f_1]$. The justification for this is the Uniformization theorem, which states that every Riemann surface is conformally equivalent to one that admits a hyperbolic metric, and this metric is uniquely determined, up to isometry, by the conformal equivalence class of the above-mentioned Riemann surface.

3.1 The Interplay Between $\mathcal{M}_g$, Γ_g and $\mathcal{T}_g$

We exhibit this interplay as follows. Fix n distinct points, ordered $p_1, \cdots, p_n$ on the surface Σ_g, and furthermore, consider triples $(\Sigma_g, (q_1, \cdots, q_i), [f])$. The definition of equivalence classes is the same as before. The space of equivalence classes is denoted by $\mathcal{T}_g^s$, the Teichmüller space, where s stands for the number of punctures. Note that the q_i are distinct, ordered points on Σ_g; $[f]$ denotes the homotopy class of f rel $\{q_i\}$, with $f(q_i) = p_i$.

We now make use of the ordering of the points p_i, in order to describe the mapping class group Γ_g^s. By its definition, it is the group of all orientation preserving homeomorphisms $\phi : \Sigma_g \to \Sigma_g$ such that $\phi(p_i) = p_i$. The mapping class group Γ_g^s acts on the Teichmüller space $\mathcal{T}_g^s$. The resulting quotient is the moduli space $\mathcal{M}_g^s$.

How does one describe the Teichmüller space using a hyperbolic surface? One option is to remove the n points p_i; this is actually taken care of by the formula

$$\Sigma_g^s = \Sigma_g - \{p_1, \cdots, p_n\}$$

to which is associated the Euler characteristic

$$\chi\left(\Sigma_g^s\right) < 0.$$

As a second step, we take a finite but complete hyperbolic area. Next, we write down a homeomorphism involving the same hyperbolic surface Σ:

$$f : \Sigma_g \to \Sigma_g^s.$$

This procedure gives the Teichmüller space. It is somewhat more of a general construction since the definition of the equivalence is exactly that in the case $n = 0$ (i.e. no ordered points).

As it stands, Teichmüller spaces classify marked complex structures on an n oriented surface of genus g with n marked points when $2g - 2 + n > 0$. As such, Teichmüller spaces carry an inherent structure of complex $(3g + 3 - n)$-manifold as well as a complete, global metric d, defined in terms of the least logarithmic distortion necessary in deforming one complex structure to another. For surfaces of genus one, these spaces are closely related to the classical modular group $\Gamma = \text{SL}(2, \mathbb{Z})$, and to the action of Γ which acts by linear transformations on the upper half plane $\mathcal{U} = T_1$; this action carries the inherently richer structure of hyperbolic plane.

The mapping class group $\text{mod}(g, n)$ acts as the generalized version of modular groups by changing the marking on a reference surface. When $n = 0$, the quotient of Teichmüller space by this action is none other than Riemann's moduli space $\mathcal{M}_g$ parametrizing bi-holomorphic equivalence classes of compact genus g Riemann surfaces.

3.2   Fenchel-Nielsen Description of Teichmüller Spaces

Fenchel and Nielsen's [1, 2] description of Teichmüller space in terms of hyperbolic coordinates is a very revealing one. It applies to Riemann surfaces of arbitrary genus and starts with the existence in the hyperbolic plane of hexagons. These hexagons are determined up to isometry by the lengths of three alternate sides. The lengths can be arbitrarily prescribed in $[0, \infty]$.

The gluing lemma allows us to take two copies of a given hexagon and glue the alternate sides to obtain a pair of pants $P = P(l_1, l_2, l_3)$. Note that P has geodesic boundary, as sketched in Figure 3.2.

According to Fenchel and Nielsen's work [1, 2], almost all hyperbolic surfaces (of finite topological type) can be obtained in this way. It is easy to see why: the parameters for the gluing are global coordinates for Teichmüller space.

To construct a surface of genus g requires $2g - 2$ pairs of pants (the Euler characteristic for a pair of pants is -1). Each pair of pants has three boundaries and as such, they can be identified in pairs. There are three times $\frac{1}{2}(2g - 2)$ gluing sites. At each site, there are two parameters, namely l_i and τ_i. In total, we have $6g - 6$ real parameters. The Fenchel-Nielsen

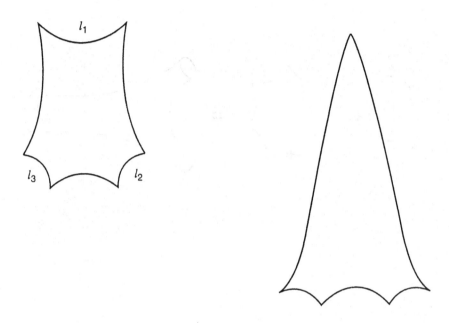

Figure 3.1: Hexagons with alternating sides in the hyperbolic plane

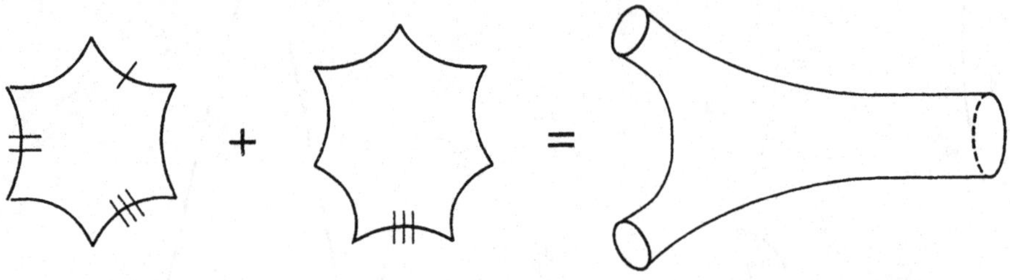

Figure 3.2: The Gluing Lemma is applied to give a pair of pants P with geodesic boundary

coordinates are thus the free parameters for this construction. The first of these parameters, l_i, denotes the length of the curve. It is essential for the gluing procedure that these curves (or sites) have the same length. As for the second parameter, τ_i, it is a twist parameter which determines the hyperbolic distance between the sites. The parameter l_i varies freely in $\mathbb{R}^+$, while τ_i varies in $\mathbb{R}$.

We wish to point out that there is, to date, no elementary connection between the Fenchel-Nielsen coordinates and the complex structure of Teichmüller spaces. Nevertheless, the expression for the Weil-Petersson Kähler form is quite simple in Fenchel-Nielsen coordinates:

$$\omega = \sum_i d\tau_i \wedge dl_i.$$

In particular, the Weil-Petersson volume form,

$$dV = \omega^{3g+3}$$

is a Euclidean volume, and the mapping class group Γ_g acts by Euclidean volume, preserving transformations.

The Fenchel-Nielsen coordinates may also be used to describe the Deligne-Mumford compactification [13] $\overline{\mathcal{M}_g^s}$ of $\mathcal{M}_g^s$ as follows. Consider a partition of Σ_g^s denoted by $\{c_i\}$; allowing some of the parameters l_i to be zero gives rise to a Riemann surface with nodes located at those c_i. $\overline{\mathcal{M}_g^s}$ is obtained from $\mathcal{M}_g^s$ by adjoining these singular surfaces. The complement has irreducible components $D_0, D_1, \cdots, D_{[g/2]}$, of real codimension two. We called D_0 a compactification locus, which is the technical term to characterize a collection of surfaces with a node at a nonseparating curve. D_i $(i > 0)$ consists of the surfaces with a node at a curve which separates Σ_g into surfaces of genus i and $g - i$. For the genus one case, the Teichmüller space is the upper half-plane H ($\tau \in H$ parametrizes the lattice with generators $\{1, \tau\}$). The resulting genus one moduli space is $H/\mathrm{PSL}\,(2; \mathbb{Z})$.

3.3   Homology of Moduli Spaces and Mapping Class Groups

The length spectrum of hyperbolic surfaces (i.e. the collection of real numbers which occur as lengths of closed geodesics) is discrete. Using this fact, one can show, without much effort, that the action of the mapping class group Γ_g^s on the Teichmüller space $\mathcal{T}_g^s$ is properly discontinuous, that is to say, that for every set $K \subset \mathcal{T}_g^s$, the collection of $\Phi \in \Gamma_g^s$ is such that, $\Phi(K) \cap K \neq \varnothing$, is finite. This means that the resulting moduli space $\mathcal{M}_g^s$ is an orbifold: each point has a neighborhood modeled on $\mathbb{R}^N$ modulo a finite group. Furthermore, since $\mathcal{M}_g^s$ is rational in K-theory, we have:

$$H_* \left(\mathcal{M}_g^s; \mathbb{Q} \right) \cong H_* \left(\Gamma_g^s; \mathbb{Q} \right). \qquad (3.2)$$

The mapping class group Γ_g^s is virtually torsion-free. To see this, let

$$\mu : \Gamma_g^s \to \mathrm{Sp}\,(2g; \mathbb{Z})$$

be the map obtained by allowing a homeomorphism Φ of Σ_g^s act on $H_1\,(\Sigma_g; \mathbb{Z})$. This gives an element of Sp since Φ preserves the intersection form on Σ_g. More precisely, the map μ fits into the exact sequence

$$1 \to T_g^s \xrightarrow{\mu} \Gamma_g^s \to \mathrm{Sp}\,(2g; \mathbb{Z}) \to 1; \qquad (3.3)$$

here, T_g^s denotes the Torelli group, a natural subgroup of the mapping class group [3, 4]. It consists of those mapping classes of diffeomorphisms of Σ_g which act trivially on its homology. This implies that the the Torelli group is in fact kernel of the (natural) surjective homomorphism

$$\Gamma_g^s \to \mathrm{Sp}\,(2g; \mathbb{Z}).$$

Looking at G_n, the full congruence subgroup of level n in $\mathrm{Sp}\,(2g; \mathbb{Z})$, we see that for $n \geq 3$, G_n is torsion free. Combining this with Johnson's result outlined in reference [3], we deduce that the Torelli group is indeed torsion-free. This, in turn, implies that the congruence subgroup $\Gamma_g^s\,[n] = \mu^{-1}\,(G_n)$ is torsion-free if $n \geq 3$ as well! Its index is the order of the finite group $\mathrm{Sp}\,(2g; \mathbb{Z}/n\,\mathbb{Z})$. The quotient, $\mathcal{T}_g^s/\Gamma_g^s\,[n] = \mathcal{M}_g^s\,[n]$, gives us back the moduli space. But this time, the quotient describes a manifold, and consequently, it follows that

$$H_* \left(\Gamma_g^s\,[n]; \mathbb{Z} \right) \cong H_* \left(\mathcal{M}_g^s\,[n]; \mathbb{Z} \right). \qquad (3.4)$$

At times, it will be necessary to compare the homology groups of Γ_g^s as we vary g and s. For s, we have the exact sequence:

$$1 \to \pi_1 \left(\Sigma_g^s \right) \to \Gamma_g^{s+1} \xrightarrow{\eta} \Gamma_g^s \to 1. \qquad (3.5)$$

The Lyndon-Hochschild-Serre Spectral Sequence (LHSSS) [5] is the most natural and readily available tool used to relate $H_*\left(\Gamma_g^s\right)$ to $H_*\left(\Gamma_g^{s+1}\right)$. In general, the LHSSS relates the cohomology of a group to that of a normal subgroup and that of the factor group. However, when we vary g there is no natural way of mapping Γ_g^s to Γ_{g+1}^s, and this is why, it becomes necessary to introduce the mapping class group of surfaces with boundaries. We postpone to Chapter 13 the crucial role played by boundary components in determining three-dimensional mapping class groups. Harer's stability theorem tells us that for $g \gg k$, $H_g(\Gamma_g^s)$ is independent of the genus g. Harer's theorem is not only an important tool for investigating mapping class groups and moduli space; it extends far beyond this proven use, particularly in quantum topology where it performs much service, for example, in the detection and cancellation of global anomalies, as we shall learn in Chapter 13.

For now, we pursue our task to give a sound mathematical basis to our knowledge of moduli spaces, Teichmüller spaces and mapping class groups.

3.4   Algebraic Structures of the Mapping Class Group

The message from formula (3.4) is simply that the homology of the moduli space and that of the mapping class group are intimately related. Thus, by investigating specific properties of the mapping class group, one may come up with a better understanding of the moduli space and/or vice versa. This intimate connection is in fact a welcome addition in our quest to better comprehend issues in 3-dimensional physics, particularly as they relate to quantum gravity and Chern-Simons-Witten theories. So we begin this section by describing a finite presentation for the mapping class group Γ_g^s.

Consider a simple closed curve $c \subset \Sigma_g^s$; the Dehn twist τ_c of c is (the isotopy class of) the homeomorphism of the Riemann surface Σ_g^s obtained by splitting along c, rotating one side 360° to the right and regluing (cf: Figure 3.3).

The proof that Γ_g^s is generated by Dehn twists on a finite number of curves is due to Dehn [7]. Humphries later determined the minimal number of generators necessary [8]. McCool [9] gave an indirect proof that Γ_g is finitely presented, but the first explicit presentation was provided by Hatcher and

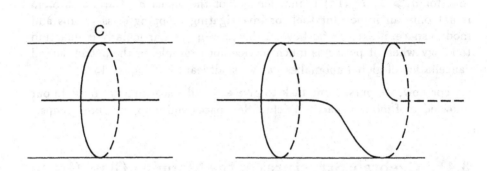

Figure 3.3:

Thurston [10]. The later was somewhat simplified by Wajnryb [11] whose presentation uses as generators:

$$\tau_i = \tau_{c_i}, \quad 1 \le i \le 2g+1,$$

and has $\binom{2g+1}{2} + 3$ relations when $g > 2$. What are these generators? The first of these are the braid relations:

$$\begin{aligned} \tau_i \tau_j &= \tau_j \tau_i & \text{if } c_i \cap c_j = \emptyset \\ \tau_i \tau_j \tau_i &= \tau_j \tau_i \tau_j & \text{if } c_i \cap c_j \ne \emptyset. \end{aligned}$$

The above presentation has an interesting algebraic consequence: it generates a group with $H_1 \cong \mathbb{Z}$. This group is normally generated by any τ_i, whereas, in actual life, $H_1 = 0$ [12, 13].

Adding the Lantern relation [14] gives a group with $H_1 = 0$. Now, the result implies that $H_2 = 0$, whereas H_2 should be $\mathbb{Z}$. In order to fix this discrepancy, we add a variant of the lantern relation, the Chinese lantern relation to get $H_2 \cong \mathbb{Z}$. It turns out that this is none other than a presentation of the mapping class group of a surface with one boundary component. To obtain a mapping class group without punctures (i.e. Γ_g^0), we add one more relation called the boundary relation. Wajnryb provides us, in reference [11], with the explicit forms of these relations. The exact sequence (3.5) is then used to complete the presentation for the group Γ_g^0.

3.5 References

[1] Abikoff, William: **The Real Analytic Theory of Teichmüller Space**, Springer-Verlag Lectures Notes in Mathematics Vol. 820 (1980).

[2] Ahlfors, L: *Some Remarks on Teichmüller Space of Riemann Surfaces*, Ann. Math. 74 (1961) 171-191.

[3] Johnson, D.: *A Survey of the Torelli Group*, Contemp. Math. 20 (1983) 165-179.

[4] Baadhio, R. A.: *Mapping Class Groups for D = 2 + 1 Quantum Gravity and Topological Quantum field Theories* Nuclear Physics B441 Nos. 1-2 (1995) 383-401.

[5] Benson, D. J. and Cohen, F. R.: **Mapping Class Groups of Low Genus and their Cohomology**, Memoirs, American Math. Society Vol 90 Number 443 (1991).

[6] Harer, J.: *Stability of the Homology of the Mapping Class Group of Orientable Surfaces*, Ann. of Math. 121 (1985) 215-249.

[7] Dehn, M.: *Die Gruppe der Abbildungsklasser*, Acta Math. 69 (1938) 81-115.

[8] Humphries, S.: *Generators for the Mapping Class Group*, in **Topology of Low Dimensional Manifolds**, Springer-Verlag Lecture Notes in Mathematics Vol. 722 (1979) pp. 44-47.

[9] McCool, J.: *Some Finitely Presented Subgroups of the Automorphism Group of a Free Group*, Journal of Algebra 35 (1975) 205-213.

[10] Hatcher, A. and Thurston, W.: *A Presentation of the Mapping Class Group of a Closed, Orientable Surface*, Topology, 19 (1980), 221-237.

[11] Wajnryb, B.: *A Simple Presentation for the Mapping Class Group of an Orientable Surface*, Israel Jour. Math. 45 (1983) 157-174.

[12] Powell, J.: *Two Theorems on the Mapping Class Group of a Surface*, Proc. Amer. Math. Society 68 (1978) 347-350.

[13] Mumford, D.: *Abelian Quotients of the Teichmüller Modular Group*, Journ. d' Analyse Math. 18 (1967) 227-244.

[14] Harer, J.: *The Second Homology Group of the Mapping Class Group of an Orientable Surface*, Invent. Math. 72 (1982) 221-239.

Chapter 4

Mapping Class Groups and Arithmetic Groups

We shall discuss, in this chapter, the close relationship between mapping class groups and arithmetic groups.

Let G be an algebraic subgroup of GL (n) defined over $\mathbb{Q}$, $G_{\mathbb{Q}}$ the group of $\mathbb{Q}$-points of G and $G_{\mathbb{Z}} = G_{\mathbb{Q}} \cap$ GL $(n; \mathbb{Z})$. A subgroup $\Gamma < G_{\mathbb{Q}}$ is called arithmetic if it is commensurable with $G_{\mathbb{Z}}$ [1]. Mapping class groups have many properties in common with arithmetic groups. The following list, due to Serre [1], gives a fairly good indication of this. Here, Γ denotes an arithmetic group: or a mapping class group

1. Γ is finitely presented;

2. Γ has only a finite number of conjugacy classes of finite subgroups;

3. Γ is residually finite;

4. Γ is virtually torsion-free;

5. For any torsion-free subgroup $\hat{\Gamma}$ of finite index in Γ, there exists a finite complex which is a $K\left(\hat{\Gamma}, 1\right)$;

6. The virtual cohomological dimension (vcd) of Γ is finite.

References for properties 1-6, when Γ is arithmetic, can be found in [1]. For the mapping class group, see [2] and [3] for property 1, [4] for property 3.

For the mapping class group, the next property on Serre's list for arithmetic groups is:

7. $H^q(\Gamma; \mathbb{Z}[\Gamma])$ is zero except for a single value of q ($q = \text{vcd}(\Gamma)$), for which it is a free $\mathbb{Z}$-module. This means that Γ is a virtual duality group, as defined by Bieri and Eckmann [5].

The following holds true for Γ arithmetic:

8. Every normal subgroup of Γ is either of finite index, or else is finite and central.

This particular property fails for the mapping class group Γ_g^s mainly because the Torelli group $T_g^s = \ker(\Gamma_g^s) \to \text{Sp}(2g; \mathbb{Z})$ is normal and is (unfortunately) neither finite nor of finite index. We will show that Γ_g^s is not arithmetic for $g \geq 3$. Actually, one extra step will show that Γ_g^s cannot be a lattice in any algebraic group G. The more general question of whether mapping class groups admit any faithful representation at all in an algebraic group remains an open problem.

4.1 Mapping Class Groups and Duality Groups

When the arithmetic group Γ is torsion-free, it follows that:

$$H^k(\Gamma; \mathbb{Z}\Gamma) \cong H_c^k(\bar{X}) \cong H_{n-k-1}(\partial\bar{X}) \cong \begin{cases} I, & k = n - d - 1 \\ 0, & \text{otherwise} \end{cases} . \quad (4.1)$$

Here, $I = H_d(\partial\bar{X})$ is a free abelian group of infinite rank commonly called a Steinberg module. Incidentally, X is the symmetric space associated to G, $\bar{X}$ denotes the Borel-Serre bordification of X, $n = \dim X$; and H_c^* is cohomology with compact supports. Γ is hence a duality group in the sense of Bieri and Eckmann [5]. This confirms the statement that for M a $\mathbb{Z}\Gamma$-module,

$$H^k(\Gamma; M) \cong H_{v-k}(\Gamma; M \otimes I); \quad (4.2)$$

where $v = n - d - 1$ is the cohomological dimension of Γ. The action of Γ extends to a properly discontinuous action on $\bar{X}$. The boundary $\partial\bar{X}$ satisfies $H_d(\partial\bar{X}) = I$. The cohomological dimension of Γ, cd(Γ), is the smallest integer v, such that there exists a $\mathbb{Z}\Gamma$-module M with $H^v(\Gamma; M) \neq 0$.

When the mapping class group Γ_g^s is a virtual duality group, a result similar to that of equation (4.2) holds. However, one has to take into account a subgroup of finite index in Γ_g^s that is torsion-free. This yields the relation

$$H^k\left(\Gamma, \mathbb{Z}\Gamma\right) \cong \begin{cases} I = H_d\left(\partial\bar{X}\right), & k = n - d - 1 \\ 0, & \text{otherwise} \end{cases} \qquad (4.3)$$

Note that n is the dimension of the Teichmüller space T_g^s, which is $6g-6+2s$; I stands for $H_d\left(\partial\bar{X}\right)$; it is a Steinberg module of $G_{\mathbb{Q}}$. Consequently,

$$H^k\left(\Gamma; M\right) \cong H_{v-k}\left(\Gamma; M \otimes I\right),$$

we have

$$v = n - d - 1 = \begin{cases} 4g - 5 & s = 0 \text{ and } g > 1 \\ 4g - 4 + s & s > 0 \text{ and } g \geq 1 \\ 1 & s = 0 \text{ and } g = 1 \\ s - 3 & s > 2 \text{ and } g = 0 \end{cases} \qquad (4.4)$$

The integer v is referred to as the virtual cohomological dimension (vcd) of the mapping class group Γ_g^s, that is, v is the cohomological dimension of any torsion-free subgroup of finite index in Γ_g^s.

4.2   Non-Arithmetic Mapping Class Groups

Assume that the group $\Gamma = \Gamma_g^s$ be an arithmetic subgroup of the linear algebraic group G. According to Borel and Serre, when n is the dimension of symmetric space X associated to G, the relation $n - \Gamma_{\mathbb{Q}}(G) = \text{vcd}(\Gamma)$ holds. The number $\Gamma_{\mathbb{Q}}(G)$ can be computed directly from Γ; it is the maximal rank of an abelian subgroup of Γ. In reference [6], this is shown to be $3g - 3 + s$. Consequently, $n = 7g - 8$ when $g > 1$, $s = 0$, and $n = 7g - 7 + 2s$ for $g > 0$. Thus, Teichmüller space cannot be identified with a symmetric space X when $s \geq 0$ and $g > 2$, or $s > 0$ and $g = 2$.

When is Γ non-arithmetic? Suppose first that G has rank one so that X is a hyperbolic space. Two elements of infinite order in the group of isometries of X then commute if and only if they have the same fixed point set on the sphere at ∞. This implies that commuting is an equivalence relation on the set of elements of infinite order in Γ. For the mapping class group, this is absurd. This is why. Simply take curves c_1 and c_2 with c_0 disjoint from c_1 and c_2, but with c_1 intersecting c_2. If τ_i denotes the Dehn twist on c_i, then τ_0 commutes with τ_1 and τ_2, but τ_1 and τ_2 do not commute.

Next, consider the case where G has rank greater than or equal to two. Property 8 of arithmetic groups on Serre's list and the existence of the Torelli group assert that G cannot be simple. Suppose then that G is semisimple; a stronger version of Property 8 says, in essence, that if Γ is an irreducible lattice in G, then, once again, any normal subgroup in G is either finite (and central) or of finite index. This means that Γ must be reducible, so there is a subgroup of finite index in Γ which is a direct product of infinite groups. An analysis of the centralizers of the elements of Γ shows that this is not possible. Finally, if G is not semisimple, we need to look at solvable subgroups of Γ. Here, a theorem by Birman et al. [6] states that every solvable subgroup of Γ is virtually abelian. Such a subgroup will not be normal in Γ, and so the map $G \rightarrow G/\mathrm{rad}\,(G)$ imbeds Γ in the semisimple group $G/\mathrm{rad}\,(G)$. This, of course, is clearly impossible.

Hence, using either method, we arrive at the conclusion that the mapping class group Γ_g^s is not an arithmetic group.

4.3   How Stable Are Mapping Class Groups?

We now return to a direct study of $H_*\left(\Gamma_g^s\right)$. A natural question to ask about the mapping class group Γ_g^s is whether its homology stabilizes as $g \rightarrow \infty$. This kind of result is known (rationally at least) for many classes of arithmetic groups, e.g. $\mathrm{SL}\,(n)$ and $\mathrm{Sp}(2n)$, as shown by Borel in reference [7]. Furthermore, the techniques to prove it are readily available [8], [9], [10]. We will combine these techniques with the results of Chapter 3 to show that, in fact, $H_k\left(\Gamma_g^s\right)$ is independent of g when $g \gg k$. Our starting point is once again the Riemann surface $\Sigma_{g,r}^s$ of genus g with r boundary components and s punctures. The mapping class group is defined to be the group of all isotopy classes of homeomorphisms of Σ_g which are the identity on $\partial\Sigma_g$ and fix the punctures individually. The isotopies in question are required to fix the boundary of Σ_g pointwise, otherwise there would be no distinction between $\Gamma_{g,r}^s$ and $\Gamma_{g,0}^{r+s}$.

Consider the following maps

$$\Phi : \Sigma_{g,r}^s \rightarrow \Sigma_{g,r+1}^s, \ r \geq 2,$$

$$\Psi : \Sigma_{g,r}^s \rightarrow \Sigma_{g+1,r-1}^s, \ r \geq 2,$$

$$\eta : \Sigma_{g,r}^s \rightarrow \Sigma_{g+1,r-2}^s, \ r \geq 2,$$

which we define as follows. For Ψ, sew a pair of pants (a copy of the surface $\Sigma_{0,3}^0$) to the regular surface $\Sigma_{g,r}^s$ along two components of its boundary. For Φ, sew a pair of pants to $\Sigma_{g,r}^s$ along one component of its boundary. Finally, for η, sew two components of $\partial \Sigma_{g,r}^s$ together. The maps Φ, Ψ and η induce maps of the mapping class group, that is to say, these are extensions of the identity on $\Sigma_{0,3}^0$ for Φ and Ψ. Let Φ_*, Ψ_*, and η_*, respectively be the maps they induce on homology . Harer has shown the following results [11].

Theorem 4.1 (Harer) $\Phi_* : H_k\left(\Gamma_{g,r}^s\right) \rightarrow H_k\left(\Gamma_{g,r+1}^s\right)$ *is an isomorphism for* $g \geq 3k - 2$

$\Psi_* : H_k\left(\Gamma_{g,r}^s\right) \rightarrow H_k\left(\Gamma_{g+1,r-1}^s\right)$ *is an isomorphism for* $g \geq 3k - 1$.

$\eta_* : H_k\left(\Gamma_{g,r}^s\right) \rightarrow H_k\left(\Gamma_{g+1,r-1}^s\right)$ *is an isomorphism for* $g \geq 3k$.

Combining these maps in various ways, one observes that Harer's theorem implies that $H_k\left(\Gamma_{g,r}^s\right)$ is independent of g and r as long as $g \geq 3k + 1$.So this pretty much answer the question of how stable mapping class groups are. For the moduli space, this is equivalent to stating that $H_k\left(\mathcal{M}_g^s; \mathbb{Q}\right)$ does not depend on g when $g \geq 3k + 1$. We defer to Chapter 13 the special role played by the mapping class group's homological stability in the detection and cancellation of global gravitational anomalies.

4.4 Computing the Cohomology of Mapping Class Groups

Our principal aim here is to construct some cohomology classes for the mapping class group Γ_g^s. We will first discuss which Betti numbers of Γ_g^s and $\mathcal{M}_g^s$

are explicitly known. Then we will give Miller's construction of a polynomial algebra on even generators in the stable cohomology of Γ_g^s [12], and in the final stage describe the unstable cohomology relations for these classes, as initially shown by Mumford [13], Morita [14] and Harris [15].

4.4.1   The First Two Betti Numbers of the Moduli Space

In Chapter 3, section 3.3, we observed that Γ_g^s has no rational cohomology above dimension $4g - 4 + s$ ($4g - 4$ when $g > 1$ and $s = 0$), while the moduli space $\mathcal{M}_g^s$ has no integral cohomology above dimension $4g - 4 + s$ when $s > 0$. Only two other Betti numbers, β_1 and β_2, are known. Mumford [13] proved that $H_1(\Gamma_g)$ is torsion of order dividing 10 ($g > 1$), and Powell [16] proved that $H_1(\Gamma_g) = 0$ when $g > 2$. The general statement is best provided by

Theorem 4.2 (Mumford-Powell)

$$H_1\left(\Gamma_g^s\right) \cong \begin{cases} 0 & \text{for } g > 2 \\ \frac{\mathbb{Z}}{10\mathbb{Z}} & \text{for } g = 2 \\ \frac{\mathbb{Z}}{12\mathbb{Z}} & \text{for } g = 1 \end{cases}. \tag{4.5}$$

Mumford defines, in reference [14], $\text{Pic}(\mathcal{M}_g^s)$, the Picard group of the moduli space. Among other things, he proved that $\text{Pic}\left(\mathcal{M}_g^s\right) \cong H^2\left(\Gamma_g^s\right)$ when $g > 1$. The following result is due to Harer and Mumford:

Theorem 4.3 (Harer-Mumford)

$$H_2\left(\Gamma_g^s\right) \cong \mathbb{Z}^{s+1} \quad g > 4.$$

A nice property of Theorem 4.2 is that it holds true rationally for Γ_3^s as well as for Γ_4^s, while $H_2(\Gamma_2^s; \mathbb{Q})$ and $H_2(\Gamma_1^s; \mathbb{Q})$ have rank s. It is unknown whether $H_2(\Gamma_g^s)$ has torsion when $2 \leq g \leq 4$.

We now give a purely topological description of the generators for $H_2\left(\Gamma_g^s\right)$. Let $\xi = \text{Diff}^+\left(\Sigma_g^s\right)$ denote the group of orientation preserving diffeomorphisms of Σ_g which fix s points, equipped with the $\mathbb{C}^\infty$ topology. The group ξ has one component for each element of the mapping class group Γ_g^s. When

$2 - 2g - s < 0$, the proof is due to Earle and Eells [18] that each of these components is contractible. This means, in essence, that the classifying space $B\xi$ is an Eilenberg-MacLane space, which we write as $K\left(\Gamma_g^s, 1\right)$. Let $\Omega_2\left(B\xi\right)$ denote the second bordism group of $B\xi$. A given element of $\Omega_2\left(B\xi\right)$ has a map $\phi : X \to B\xi$ for representative, where X is a closed oriented surface. Bordism groups and homology groups agree in low dimensions; in particular, we have

$$H_2\left(\Gamma_g^s\right) \cong H_2\left(B\xi\right) \cong \Omega_2\left(B\xi\right). \qquad (4.6)$$

With these descriptions in hand, we now proceed to define $H_2\left(\Gamma_g^s\right) \to \mathbb{Z}^{s+1}$. Each $X \in H_2\left(\Gamma_g^s\right)$ gives rise to a fiber bundle $\Sigma_g \to W^4 \to X$. Such a bundle possesses s canonical sections $\sigma_1, \cdots, \sigma_s$; the invariants associated to X are (signature $W/4$) and the self-intersection numbers $[\sigma_i(X)]^2$. Note that the signature of the 4-manifold W^4 is always divisible by four; consequently these invariants are all integers. According to Theorem 4.2, the map $H_2\left(\Gamma_g^s\right) \to \mathbb{Z}^{s+1}$ is an isomorphism.

4.4.2 Miller's Polynomial Algebra

In this subsection, we use ξ to denote $\text{Diff}^+\left(\Sigma_g^0\right)$ (instead of $\text{Diff}^+\left(\Sigma_g^s\right)$ in the previous subsection). We also consider a new ingredient, the universal Σ_g-bundle $p : E \to B\xi$. Define τ to be the bundle of tangents to the fibers of p and set $\omega = c_1(\tau)$. This yields the relation

$$\kappa_i = p_* \omega^{i+1};$$

$\kappa_i \in H^2\left(B\xi\right) \cong H^{2i}\left(\Gamma_g^0\right).$

Mumford [13] showed that $f^*\left(\kappa_i\right)$ may be regarded as a Chow cohomology class in $A^i\left(B\right) \otimes \mathbb{Q}$, and Miller proved:

Theorem 4.4 *Let $\mathbb{Q}\left[\kappa_1, \kappa_2, \cdots\right]$ be the polynomial algebra in the κ_i. The resulting map*

$$\mathbb{Q}\left[\kappa_1, \kappa_2, \cdots\right] \to H^*\left(\Gamma_g^0; \mathbb{Q}\right)$$

is injective in dimensions less than $g/3$.

To prove this theorem, Miller first looked at the Hopf algebra

$$\mathcal{A} = \lim H_* \left(\Gamma_{g,1}; \mathbb{Q}\right)$$

and showed that each of the κ_i vanish. It therefore suffices to construct explicit homology classes $X_n \in H_{2n}\left(\xi; \mathbb{Q}\right)$ such that $\langle \kappa_n, X_n \rangle \neq 0$. The theorem then follows from Harer's stability theorem.

4.4.3   Comparison With $H^*\left(\mathrm{Sp}(2g; \mathbb{Z}); \mathbb{Q}\right)$

In reference [7], Borel found the stable cohomology of $\mathrm{Sp}(2g; \mathbb{Z})$ with $\mathbb{Q}$-coefficients; it is a polynomial algebra on generators in dimensions $4k + 2$, $k \geq 0$. Miller considered the problem of computing the map induced on cohomology by

$$\mu : \Gamma_g \rightarrow \mathrm{Sp}(2g; \mathbb{Z}).$$

The map μ induces a map $B\,\mathrm{Diff}^+ \left(\Sigma_g\right) \rightarrow B\,Sp(2g; \mathbb{Z})$. Using the inclusion $\mathrm{Sp}(2g; \mathbb{Z}) \subset \mathrm{Sp}(2g; \mathbb{R})$, and the fact that $U(g)$ is the maximal compact subgroup of $\mathrm{Sp}(2g; \mathbb{R})$, we have the map

$$\bar{\mu} : B\,\mathrm{Diff}^+ \left(\Sigma_g\right) \rightarrow B\,U(g).$$

Let ν denote the universal bundle over BU. The characteristic class $s_n\left(\nu\right)$, i.e. the polynomials in the Chern classes of ν corresponding to $\sum_g t_j^n$, vanish. Miller [12], Mumford [13] and Morita [14] have independently proven that

$$\bar{\mu}^\star \left(s_n\left(\nu\right)\right) = \begin{cases} 0, & n \text{ even} \\ (-1)^{n+1/2}\, B_n, & n \text{ odd} \end{cases} ; \tag{4.7}$$

where B_n is the n-th Bernoulli number.

4.4.4   Relations Among the κ_i

Mumford, in reference [13], gave an algebraic construction of the classes κ_i and exhibited some relations among them. These and other relations (along with Miller's polynomial algebra) were rediscovered by Morita [14] who gave a topological description of them. Harris [15], while considering the problem of determining when certain linear combinations of divisor classes are ample and/or effective, also found relations among the classes in Pic $\left(\mathcal{M}_g^1\right) \otimes \mathbb{Q}$. The purpose of this subsection is to briefly describe these relations.

Let η denote the g-dimensional complex vector bundle over $B\xi$ induced by $\bar{\mu}$, and furthermore let $s_i(\eta)$ be the pull-back of the characteristic class $s_i(\nu)$. As a real bundle, η is determined by the map $\Gamma_g \to \mathrm{Sp}(2g;\mathbb{R})$; it is a flat bundle. This implies that all its Pontrjagin classes vanish, that is, $s_{2i}(\eta) = 0$. This generates a relation in $H^{4i}\left(\Gamma_g^0;\mathbb{Q}\right)$. Mumford and Morita have independently observed that $c_k(\ker \phi) = 0$ for $k \geq g$. This gives

$$\sum_{j=0}^{g} \kappa^{k-j} c_j = 0; \tag{4.8}$$

where $\kappa = \pi^*(\kappa_1) \in H^2(E)$ and c_j stands for the pull-back to $H^{2j}(E)$ of the j-th Chern class of η. We write ξ as $\pi : E \to B\Lambda$, the universal bundle over $B\Lambda$. Since E is an element of $K\left(\Gamma_g^1, 1\right)$ this means that equation (4.8) is actually generated by $H^*\left(\Gamma_g^1;\mathbb{Q}\right)$. Applying π_* to formula (4.8) gives

$$\sum_{j=0}^{q} \kappa_{k-1-j} \pi_*(c_j) = 0, \tag{4.9}$$

a relation in $H^*(\Gamma_g;\mathbb{Q})$. Mumford showed that these relations imply that κ_i is a polynomial in $\kappa_1, \cdots, \kappa_{g-2}$ for all $i \geq g-1$.

In [15], Harris considers the following question. Let ω be the first Chern class c_1 of the relative dualizing sheaf, and furthermore, let $\kappa = \pi^*\pi_*\omega^2 \in H^2(\mathcal{M}_g;\mathbb{Q}) \cong \mathrm{Pic}(\mathcal{M}_g) \otimes \mathbb{Q}$. Is the relation

$$\kappa_1^2 > (2g-2)\kappa_2 \tag{4.10}$$

valid? When $B\xi$ is at least of dimension two, the relation is known to hold true. On the basis of this observation, Harris has derived the relation:

$$(4g(g-1)\omega - \kappa)^{g+1} = 0. \tag{4.11}$$

4.5 Homology of the Deligne-Mumford' Moduli Space $\overline{\mathcal{M}}_g^s$

The Mayer-Vietoris spectral sequence easily relates $H_* \left(\mathcal{M}_g^s \right)$ to $H_* \left(\overline{\mathcal{M}}_g^s \right)$. For instance $\beta_1 \left(\mathcal{M}_g^s \right) = 0$ implies $\beta_1 \left(\overline{\mathcal{M}}_g^s \right) = 0$ and from previous sections, we can see that $\beta_2 \left(\mathcal{M}_g \right) = 1$. This implies $\beta_2 \left(\overline{\mathcal{M}}_g \right) = 2 + [g/2]$. The fact that κ_1 is symplectic on $\overline{\mathcal{M}}_g$ shows that $\kappa_1^{3g-3} \neq 0$. Wolpert uses intersections of the D_i (i.e. the compactification locus defined in Chapter 3, section 3.2.) to show that $\beta_{2k} \left(\overline{\mathcal{M}}_g \right) \geq \frac{1}{2} \begin{pmatrix} g-1 \\ k \end{pmatrix}$. The group $\mathrm{Sp}\left(2g; \mathbb{Z} \right)$ acts on the Siegel upper half space S_g by the formula

$$Z \cdot M = (ZC + D)^{-1} \cdot (ZA + B),$$

where $M = \begin{pmatrix} A & B \\ C & D \end{pmatrix}$.

The quotient $S_g / \mathrm{Sp}\left(2g; \mathbb{Z} \right)$ is called the Siegel moduli space and may be identified with $\mathcal{A}_g$, the space of (polarized) abelian varieties. The action of $\mathrm{Sp}\left(2g; \mathbb{Z} \right)$ is properly discontinuous. Consequently,

$$H^* \left(\mathcal{A}_g; \mathbb{Q} \right) \cong H^* \left(\mathrm{Sp}(2g; \mathbb{Z}); \mathbb{Q} \right).$$

To a given Riemann surface, we associate its Jacobian; this defines a mapping

$$J : \mathcal{M}_g \rightarrow \mathcal{A}_g,$$

which we refer to as the period mapping. A classical theorem of Torelli tells us that J is an embedding. Rationally, the maps J^* and μ^* are the same on cohomology. Next, let $\overline{\mathcal{A}}_g$ denote the Satake compactification of $\mathcal{A}_g$; it is obtained in a manner similar to $\overline{\mathcal{M}}_g$ by adding on copies of $\mathcal{A}_k$, $k < n$ at ∞. The map J extends to

$$J : \overline{\mathcal{M}}_g \rightarrow \overline{\mathcal{A}}_g.$$

Charney and Lee [19] have shown the following results, which we summarize as follows:

Theorem 4.5 (1) For $g \geq i + 1$, $H^i \left(\overline{\mathcal{A}}_g; \mathbb{Q} \right) \cong H^i \left(\lim_g \overline{\mathcal{A}}_g; \mathbb{Q} \right)$.

(2) $H^\star \left(\lim \overline{\mathcal{A}_g}; \mathbb{Q} \right) \cong \mathbb{Q}[x_2, x_6, \cdots] \otimes \mathbb{Q}[y_6, y_{10}, \cdots]$, *where the* x_i *live in* $H^\star(\mathcal{A}_g; \mathbb{Q})$.

(3) $\ker\left(\overline{J^\star}\right) = <y_6, y_{10}, \cdots>$, *that is,* $\mathrm{image}(\overline{J^\star}) = \mathrm{image}(\mu^\star)$.

This result is proven by decomposing $\overline{\mathcal{A}_g}$ as a union of simplicial $K(\pi, 1)$'s, relating the pieces to K-theory to compute the cohomology.

4.6 Torsion in the Cohomology of the Mapping Class Group

For completeness we will present here what is known about torsion in the cohomology of the mapping class group $H^\star(\Gamma_g^0; \mathbb{Z})$. The relationship between torsion and global anomalies is analyzed in Chapter 13. According to Harer's stability theorem, $H^\star\left(\Gamma_g^0; \mathbb{Z}\right) \cong H^\star\left(\Gamma_{g,1}^0; \mathbb{Z}\right)$. Since $\Gamma_{g,1}^0$ is torsion-free, we deduce that $H^\star(\Gamma_{g,1}^0; \mathbb{Z}) \cong H^\star(\mathcal{M}_{g,1}^0; \mathbb{Z})$. Here, $\mathcal{M}_{g,1}^0$ is the moduli space of triples (Σ_g, p, v) with Σ_g a Riemann surface, p a point in Σ_g, and v a unit tangent vector to Σ_g at p.

Note that we only have $H^\star(\Gamma_g; \mathbb{Q}) \cong H^\star(\mathcal{M}_g; \mathbb{Q})$. It is not known whether the torsion classes we shall describe lie in $\mathcal{M}_g$. Outside of low-dimensional phenomena, the first in $H^\star(\Gamma_g)$ seems to be due to Charney and Lee [20]. As a special case of their work on classifying spaces of Hodge structures, they prove that for every odd prime p, there exists a p-torsion class in $H^{2p-2}\left(\Gamma_{(p-1)/2}^0; \mathbb{Z}\right)$. This result was strengthened by Glover and Mislin [21] who showed that the stable cohomology group

$$H^{4k}(\Gamma_g; \mathbb{Z}),$$

$(g \gg k)$ contains an element of order E_{2k}, i.e. the denominator of $B_{2k}/2k$, where B_{2k} is the $2k$-th Bernoulli number.

For prime p, the number E_{2k} is divisible by p^α if and only if $p^{\alpha-1}(p-1)$ divides $2k$. This means that for a given prime p and odd with $\alpha \geq 1, j \geq 1$, $H^{2jp^\alpha(p-1)}\left(\Gamma_g^0; \mathbb{Z}\right)$ contains an element of order p^α.

4.7   References

[1] Serre, J.-P.: *Arithmetic Groups* in **Homological Group Theory** Ed. Wall, C. T. C., Cambridge University Press 1979.

[2] Wajnryb, B.: *A Simple Presentation for the Mapping Class Group of an Orientable Surface*, Proc. Amer. Math. Society 68 (1978) 347-530.

[3] Hatcher, A. and Thurston, W.: *A Presentation of the Mapping Class Group of a Closed, Orientable Surface*, Topology 19 (1980) 221-237.

[4] McCool, J.: *Some Finitely Presented Subgroups of the Automorphism Group of a Free Group*, Journal of Algebra 35 (1975) 205-213.

[5] Bieri, R. and Eckmann, B.: *Groups With Homological Duality Generalizing Poincaré Duality*, Inv. Math. 20 (1973) 103-124.

[6] Birman, J., Lubotzky, A. and McCarthy, J.: *Abelian and Solvable Subgroups of the Mapping Class Group*, Duke Math. Journal 50 (1983) 1107-1120.

[7] Borel, A.: *Cohomologie des Sous-Groupes Discrets et Representations de Groupes Semisimple*, Asterique 32-33 (1976) 73-112.

[8] Charney, R.: *Homology Stability for $GL(n)$ of a Dedekind Domain*, Inv. Math. 27 (1980) 1-17.

[9] Vogtmann, K.: *Spherical Posets and Homology Stability for $O_{n,n}$*, Topology 20 (1981) 119-132.

[10] Wagoner, J.: *Stability for the Homology of the General Linear Group of a Local Ring*, Topology 15 (1976) 417-423.

[11] Harer, J.: *Stability of the Homology of the Mapping Class Group of Orientable Surfaces*, Ann. of Math. 121 (1985) 215-249.

[12] Miller, E.: *The Homology of the Mapping Class Group*, Journal Diff. Geometry 24 (1986) 1-14.

[13] Mumford, D.: *Abelian Quotients of the Teichmüller Modular Group*, Journal d' Analyse Math. 18 (1967) 227-244.

[14] Morita, S.: *Casson's Invariant for Homology 3-Spheres and Characteristic Classes of Surface Bundles I*, Topology 38 (1989) 305-323.

[15] Harris, J.: *Families of Smooth Curves*, Preprint 1984.

[16] Powell, J.: *Two Theorems on the Mapping Class Group of a Surface*, Proc. Amer. Math. Society 68 (1978) 347-350.

[17] Harer, J.: *The Second Homology Group of the Mapping Class Group of an Orientable Surface*, Inv. Math. 72 (1982) 221-239.

[18] Earle, C. and Eells, J.: *The Diffeomorphism Group of a Compact Riemann Surface*, Bull. Amer. Math. Society 73 (1967) 371-380.

[19] Charney, R. and Lee, R.: *Cohomology of the Satake Compactification*, Topology 22 (1983) 389-423.

[20] Charney, R. and Lee, R.: *Characteristic Classes for the Classifying Space of Hodge Structures*, preprint 1984.

[21] Glover, H. and Mislin, G.: *Torsion in the Mapping Class Group and its Cohomology*, Journal Pure and Appl. Algebra 44 (1987) 177-189.

Chapter 5

Weil-Petersson Geometry of Teichmüller Spaces

We have seen from previous chapters that the Teichmüller space T_g^s classifies marked complex structures on an s-pointed genus g surface when $2g-2+s > 0$; as such, Teichmüller spaces carry intrinsically, a structure of complex $(3g - s + 3)$-manifold and a complete global metric d defined by Teichmüller space in terms of the least (logarithmic) distortion necessary to deform one complex structure to another. When the surface has genus one, these spaces are closely related to the classical modular group $\Gamma_1 = \mathrm{SL}(2, \mathbb{Z})$ and the action of Γ_1 (by fractional linear transformations) on the upper half-plane $\mathcal{U}$ is none other than T_g^s, which carries the inherently richer geometric structure of a hyperbolic plane. In previous chapters we have also shown that the mapping class group $\Gamma_g^s \bmod_{g,s}$ of the (pointed) surface acts as the general version of a by changing the marking on a reference surface. For $s = 0$, the quotient of Teichmüller space by this action is the Riemann moduli space $\mathcal{M}_g$, parametrizing bi-holomorphic equivalence classes of compact genus g. Teichmüller deformations are important ingredients in string theory, particularly in quantization and multiloop amplitude computations [1].

In this chapter, we will describe some results of Wolpert on the geometry of Teichmüller spaces [2, 3, 4, 5]. A whole other chapter is devoted to the quantization of Chern-Simons-Witten theories. Namely, in Chapter 8 we will make heavy use of the techniques developed here. Our starting point

will be to determine how much of the formal geometry of a symmetric space
a Teichmüller space has. The metric we shall study on $\mathcal{T}_g^s$ is the Weil-
Petersson metric [2, 5]; it is Kähler and we will see that its Hermitian and
symplectic geometry arise from the hyperbolic geometry of the surface. The
metric in question, it should be noted, is also invariant under the action of
the mapping class group Γ_g^s; on the moduli space $\mathcal{M}_g^s$, however, it is not
a complete metric; rather, it admits a continuous extension to the Deligne-
Mumford compactification $\overline{\mathcal{M}}_g^s$ [6]. The corresponding Kähler form ω_{WP}
extends to $\overline{\omega}_{WP}$ on $\overline{\mathcal{M}}_g^s$. We will show how Wolpert uses $\overline{\omega}_{WP}$ to give an
analytic proof that $\overline{\mathcal{M}}_g^s$ is projective.

5.1 Symplectic Geometry of the Weil-Petersson Form

We begin with a definition of the Weil-Petersson metric on $\mathcal{T}_g^s$. Let Σ_g denote
a Riemann surface of genus g and let λ be the hyperbolic line element on
Σ_g. Teichmüller space is a complex manifold and the holomorphic cotangent
space at Σ_g may be identified with $\Phi(\Sigma_g)$, the space of integrable holomorphic quadratic differentials on Σ_g (i.e. tensors of the type $dz \otimes dz$). For
$\phi, \psi \in \Phi(\Sigma_g)$, the corresponding Hermitian product,

$$\langle \phi, \psi \rangle = \frac{1}{2} \int_{\Sigma_g} \phi\psi\lambda^{-2}$$

defines the Weil-Petersson metric. This metric is Kähler; its corresponding
Kähler form is denoted ω_{WP}. We now give Wolpert's formula for ω_{WP} in
terms of the Fenchel-Nielsen coordinates.

5.1.1 Weil-Petersson Metric in Fenchel-Nielsen Coordinates

The surface Σ_g possesses a maximal curve system which we write as $C = \{C_1, \ldots, C_n\}$; the corresponding Fenchel-Nielsen coordinates for the Teichmüller
space $\mathcal{T}_g^s$ are labelled by (τ_i, l_i). The Weil-Petersson metric is then, by defi-
nition,

$$\omega_{WP} = \sum_i dl_i \wedge d\tau_i. \tag{5.1}$$

Several things about equation (5.1) are surprising. First of all, the Weil-
Petersson metric is Kähler while the Fenchel-Nielsen coordinates are only

real analytic. In view of this, the simplicity of the formula is therefore un-
expected. Secondly, the Weil-Petersson metric is invariant under the action
of the mapping class group Γ_g^s; consequently, $\sum_i dl_i \wedge d\tau_i$ must also be invariant. Actually, (5.1) says more since it shows that its right hand side is
independent of the curve system C. By contrast, the change of coordinates
from one curve system to another can be quite complicated.

Formula (5.1) is actually a consequence of the duality formula, which,
we may recall, was put to use in Chapter 4, section 4.1. To state this more
precisely, we must first define the Fenchel-Nielsen twist vector field t_C. Let
$(X, [f])$ represent a point in Teichmüller space with X hyperbolic. Furthermore, let α be the closed geodesic on X representing the free homotopy class
$f^{-1}(C)$, where $C \subset \Sigma_g$ is a nontrivial simple closed curve. We cut X along
α, rotate one side of the cut and then reglue the sides. The hyperbolic structure in the complement of the cut extends naturally to a hyperbolic structure
on the new surface. Varying the amount of rotation gives a flow on $\mathcal{T}_g^s$ and
the twist vector field adds to this flow. In what follows, we will always nor-
malize t_C so that the hyperbolic displacement of two points on opposite sides
of the geodesic α increases at unit speed (thus, for example, a full rotation
about α occurs at time $t = $ length of α.)

We introduce at this point $H(\Sigma_g)$, the space of harmonic Beltrami dif-
ferentials on Σ_g. An element $\mu \in H(\Sigma_g)$ is a tensor of type $\frac{\partial}{\partial z} \wedge d\bar{z}$ and is
harmonic with respect to the Laplace-Beltrami operator for the hyperbolic
metric on Σ_g. The holomorphic tangent space at Σ_g may be identified with
$H(\Sigma_g)$ and the Weil-Petersson metric on $\mathcal{T}_g^s$ has the dual expression

$$\langle \mu, \eta \rangle = \int_{\Sigma_g} \mu \bar{\nu} \lambda^2;$$

where $\mu, \nu \in H(\Sigma_g)$ and λ is a hyperbolic line element. The Riemannian
structure underlying $<,>$ is of course given by the symmetric tensor

$$g(\mu, \nu) = 2\text{Re} < \mu, \nu >,$$

and the Weil-Petersson Kähler form is defined by the equation

$$\omega_{WP}(\mu, \nu) = g(J\mu, \nu),$$

where J is the complex structure on the Teichmüller space. This means that
$\omega_{WP}(t_C,)$ is the Riemannian dual of Jt_C.

As a next step, we write down formulas for t_C and dl_C in terms of the Poincaré series. Let $X = \mathbb{H}/\Gamma$ and let α be the simple closed geodesic representing C in X. We use $A = \begin{pmatrix} a & b \\ c & d \end{pmatrix}$ to represent α in $\Gamma < PSL(2;\mathbb{R})$ and define $\Omega_A(\zeta) = (\mathrm{tr}^2 A - 4)(\zeta^2 + (d-a)\zeta - b)^{-2}$. Let $<A>$ denote the infinite cyclic group generated by A; this yields

$$\theta_C = \sum_{B \in \Gamma / <A>} \Omega_{B^{-1}AB}. \qquad (5.2)$$

This is a relative Poincaré series and it converges uniformly and absolutely on compact sets. A formula of Gardiner [7] expresses dl_C in terms of the relation (5.2); namely given the tangent vector $\mu \in H(X)$, we can write

$$\mathrm{Re}\,(dl_C(\mu)) = \frac{2}{\pi} \int_X \mu\, \theta_C. \qquad (5.3)$$

There is more to this. In particular, we may write $-dl_C$ as $-\frac{2}{\pi}\theta_C$. On the other hand, Wolpert [2] uses the Bers embedding (of the Teichmüller space into the vector space of Γ-invariant holomorphic quadratic differentials) to show that

$$t_C = \frac{i}{\pi}(\mathrm{Im}\,z)^2\, \bar{\theta}_C.$$

It therefore follows that $Jt_C = -\frac{1}{\pi}(\mathrm{Im}\,z)^2\, \bar{\theta}_C$, and consequently $-dl_C$ can be viewed as the Riemannian dual to Jt_C. Combining these with the duality formula gives the final form for the Weil-Petersson metric

$$\omega_{WP}(t_C) = -dl_C. \qquad (5.4)$$

A few comments are in order. Equation (5.1) could actually have been derived from (5.4). Here is how. Note that $\{C_i\}$ is a partition of Σ_g giving the Fenchel-Nielsen coordinates $\{\tau_i, l_i\}$, as shown above. Thus, the twist vector fields t_{C_i} are just the coordinate vector field $\frac{\partial}{\partial \tau_i}$. The duality formula implies that ω_{WP} is invariant under any twist flow. Putting this together with the fact that the coordinate vector fields $\{\frac{\partial}{\partial \tau_i}, \frac{\partial}{\partial l_i}\}$ commute, one observes that the coefficients of ω_{WP} in the basis $\{d\tau_i \wedge d\tau_j, dl_i \wedge d\tau_j, dl_i \wedge dl_j\}$ are independent of τ_i. We can evaluate ω_{WP} at X by using the partition $\{C_i\}(\rho(\alpha_i)) = \alpha_i$ where $\{\alpha_i\}$ are the geodesics representing $\{C_i\}$ on X, and ρ is the orientation reversing isometry. The functions l_{C_i} are invariant under ρ since the length of a given curve does not depend on the orientation of the surface. On the other hand, the twist parameter τ_i does depend on this orientation because right and left are reversed by ρ. One may make this precise by showing

$$\rho^* \, dl_{C_i} = dl_{C_i}$$
$$\rho^* \, d\tau_{C_j} = -d\tau_{C_j} + \frac{n_j}{2} \, dl_{C_j}$$

for some integers n_j. Since ρ corresponds to an element of the mapping class group which acts anti-holomorphically we have

$$\rho^* \, \omega_{WP} = -\omega_{WP}.$$

The coefficients of $d\tau_i \wedge d\tau_j$ and $dl_i \wedge dl_j$ are even relative to a ρ substitution as long as the Weil-Petersson metric, ω_{WP}, is odd. This translates into the fact that these coefficients are identically zero. It is worth pointing out that the coefficient of $dl_i \wedge d\tau_j$ is the Kronecker delta δ_{ij}, and that this follows directly from equation (5.1).

We now take a closer look at the implications of (5.1). The first and foremost consequence is that the vector fields t_C are Hamiltonian for ω_{WP}; in other words, the Lie derivative $L_{t_C} \omega_{WP}$ vanishes. This can be seen from the rather general formula

$$L_X \, \omega_{WP} = (d\omega_{WP})(X, \, , \,) + d(\omega_{WP}(X, \,)),$$

the fact that ω_{WP} is Kähler (i.e. $d\omega_{WP} = 0$), and the duality formula. Hence, the Weil-Petersson metric and the vector fields t_C define a symplectic geometry on $\mathcal{M}_g^s$ and $\mathcal{T}_g^s$. Later, we will see that ω_{WP} extends smoothly to $\overline{\mathcal{M}}_g^s$ where it remains symplectic. By drawing an analogy with symmetric spaces, we may use ω_{WP} to define a Lie algebra as follows. Take the vector space of all vector fields X on the Teichmüller space $\mathcal{T}_g^s$ such that $L_X \, \omega_{WP} = 0$. It can then be shown that the resulting associated algebra is generated over the $\mathbb{C}^\infty$ functions by the L_{t_C}; this algebra is, however, infinite dimensional.

A fundamental idea generated by equations (5.1) and (5.4) is that the hyperbolic geometry on the surface is reflected in the symplectic geometry of Teichmüller space. There are three main formulas for this. These are the cosine formula, the sine-length formula and the Lie bracket formula.

5.1.2   The Cosine Formula

Consider C_1 and C_2, two curves in Σ_g at $X \in \mathcal{T}_g^s$. The cosine formula is simply

$$t_{C_1} l_{C_2} = \omega_{WP} (t_{C_1}, t_{C_2}) = \sum_{p \in \alpha \sharp \beta} \cos \theta_p, \qquad (5.5)$$

where α and β are the geodesics in X representing C_1 and C_2, $\alpha \sharp \beta = \alpha \cap \beta$ unless of course $\alpha = \beta$, in which case it is empty, and θ_p denotes the angle between α and β. Here, $t_{C_1} l_{C_2}$ is taken to mean the Lie derivative of l_{C_2} by the vector field t_{C_1}.

5.1.3   The Sine-Length Formula

The formula reads

$$t_{C_0} t_{C_1} l_{C_2} = \sum_{(p,q) \in \alpha \sharp \gamma \times \beta \sharp \gamma} \frac{e^{m_1} + e^{m_2}}{2(e^{l_\gamma} - 1)} \sin \theta_p \cdot \sin \theta_q$$
$$- \sum_{(r,s) \in \alpha \sharp \beta \times \beta \sharp \gamma} \frac{e^{n_1} + e^{n_2}}{2(e^{l_\gamma} - 1)} \sin \theta_r \sin \theta_s, \qquad (5.6)$$

where the two possible routes from p to q along γ have lengths m_1 and m_2, and the two routes from r to s along β have the lengths n_1 and n_2. As with the cosine formula, the C_i, $i = 0, 1, 2, \cdots$, denotes curve in Σ_g while α, β and γ represent C_0, C_1 and C_2, respectively, in $X \in \mathcal{T}_g^s$.

5.1.4   The Lie Bracket Formula

As a first step, we need to renormalize the vector field t_C. This is done by setting $T_C = 4 (\sinh l_C/2) t_C$. Then, with the notation as above, we write

$$[T_{C_1}, T_{C_2}] = \sum_{p \in \alpha \sharp \beta} T_{\alpha_p \beta^+} - T_{\alpha_p \beta^-}, \qquad (5.7)$$

where $\alpha_p \beta^+$ and $\alpha_p \beta^-$ are curves in Σ_g. Formula (5.7) suggests that a Lie algebra over the integers might be constructed as follows. Let $\hat{\pi}$ denote the conjugacy classes of the fundamental group of the surface Σ_g, and furthermore, let $\mathbf{Z}\hat{\pi}$ denote the free abelian group on the elements of $\hat{\pi}$. Fixing a hyperbolic metric on Σ_g, we see that each element of $\hat{\pi}$ is represented by a unique, closed geodesic. Goldman [8] defines a bracket on $\mathbb{Z}\hat{\pi}$ using the formula

$$[\alpha, \beta] = \sum_{p \in \alpha \sharp \beta} \alpha_p \beta^+ - \alpha_p \beta^-.$$

The proof that $[\,,\,]$ is indeed a Lie bracket on $\mathbb{Z}\hat{\pi}$ is purely topological.

There is, unfortunately, a drawback to all of these definitions in that they give rise to infinite dimensional Lie algebras, and there is no indication that they contain any interesting finite dimensional subalgebras. We will not be discouraged, however, and we shall continue the study of the Weil-Petersson geometry with an eye toward the formal geometry of symmetric spaces. In the following section, we dig more deeply into the interplay between the hyperbolic geometry of Σ_g and that of the symplectic geometry of Teichmüller spaces.

5.2   Symplectic Geometry of Teichmüller Spaces

Thurston introduced the notion of a random geodesic on a hyperbolic surface X. He showed that the corresponding length function has its minimum at X and then used this concept to define a metric on Teichmüller space. Wolpert [5] is credited with the proof that this metric is actually the Weil-Petersson metric. This new interpretation of the metric proved to be quite valuable, for it gives the following remarkable formula for the complex structure on the Teichmüller space:

$$J t_C = 3\pi \, (g - 1) \lim_j \frac{[t_C, t_{\beta_j}]}{l_{\beta_j}}, \qquad (5.8)$$

where the β_j are the random geodesics.

Definition 5.1 *Definition (Random Geodesics)*

We begin with the X and write $\beta_j \subset X$ as a sequence of closed geodesics. If $T_1 X$ is the unit tangent bundle to X, then each β_j has a unique lift of $\tilde{\beta}_j$ to $T_1 X$. Consider now a uniform distribution of $\{\beta_j\}$ within X. This yields the relation

$$\lim_j \frac{l\left(\tilde{\beta}_j \cap U\right)}{l\left(\tilde{\beta}_j\right)} = \frac{\mathrm{vol}\,(U)}{\mathrm{vol}\,(T_1 X)}.$$

$U \subset T_1X$ and we have identified T_1X with $T_1 \mathbb{H}^2/\Gamma_2$, where $X = \mathbb{H}^2/\Gamma$. The volume is computed via the isomorphism $T_1\mathbb{H}^2 \cong \mathrm{PSL}(2,\mathbb{R})$. We now write the Weil-Petersson metric as $<,>_T$ instead of ω_{WP} as in previous sections. For non-trivial simple closed curves C_1, C_2 we have:

$$\langle t_{C_1}, t_{C_2} \rangle = 3\pi \, (g-1) \lim_j \frac{t_{C_1} \, t_{C_2} \, l_{\beta_j}}{l_{\beta_j}}. \tag{5.9}$$

(Compare with equation (5.8) where $\{\beta_j\}$ is uniformly distributed on X.) In the formula (5.9) we are writing l_{β_j} when we really mean $l_{f(\beta_j)}$, where $f : X \to \Sigma_g$ is the marking on X. Note that

$$< t_{C_1}, t_{C_2} >_T = \lim_j \frac{1}{l_{\beta_j}} t_{C_1} \, t_{C_2} \, l_{\beta_j}$$

is symmetric. Here is why. Since $[t_{C_1}, t_{C_2}]$ is a tangent vector for C_1 and C_2, it follows that $\lim_j \frac{1}{l_{\beta_j}} V \, l_{\beta_j} = 0$ for the tangent vector V. Since the twist vector fields span the tangent space to the Teichmüller space $\mathcal{T}_g^s$, $\lim_j \frac{1}{l_{\beta_j}} t_C \, l_{\beta_j}$ vanishes. But the β_j are uniformly distributed in T_1X, which means that for a geodesic $\alpha \subset X$ the limit of each intersection of α with β_j of angle θ is accompanied by another intersection of angle $\pi - \theta$. Applying the cosine formula completes the argument.

Another point of interest is the fact that the Weil-Petersson metric $<,>_T$ is constructed naturally with respect to the $\mathrm{PSL}\,(2,\mathbb{R})$ geometry on the unit tangent bundle T_1X. To make use of this, Wolpert relates the existence of tensors k_1 and k_2 on X to two holomorphic quadratic differentials ϕ, ψ on X which, as a prerequisite, are even continuous in complex coordinates on the Deligne-Mumford moduli space $\overline{\mathcal{M}}_g^s$. Wolpert's approach consists of three parts: the extension of the Kähler form to the Deligne-Mumford moduli, the rationality of the extension, and of course, the positivity of the resulting line bundle. They are analyzed below.

5.2.1 Extension of the Weil-Petersson Metric to Deligne-Mumford Moduli Space

Masur [6] was the first to consider the extension of ω_{WP} to $\overline{\mathcal{M}}_g^s$. At a generic point of $D = \overline{\mathcal{M}}_g^s - \mathcal{M}_g^s$ with normal coordinates z, he gave the formula

$$\bar{\omega}_{WP} \sim \frac{(i/2)\, dz \wedge d\bar{z}}{|z|^2 \left(log\, 1/|z|\right)^3}, \tag{5.10}$$

thereby showing that the extension is not continuous in the complex coordinate domain. By contrast, since the Fenchel-Nielsen coordinates on $\overline{\mathcal{M}}_g^s$ are obtained simply by allowing the l_i to be zero, the formula $\omega_{WP} = \sum dl_i \wedge d\tau_i$ shows that the Weil-Petersson metric extends smoothly in Fenchel-Nielsen coordinates. This means that the complex structure $\overline{\mathcal{M}}_g^C$ and the real analytic structure $\overline{\mathcal{M}}_g^{FN}$ are not subordinate to a common smooth structure on $\overline{\mathcal{M}}_g^s$. Nevertheless, in [5], it is proved that the identity map $\overline{\mathcal{M}}_g^{FN} \to \overline{\mathcal{M}}_g^C$ is Lipschitz continuous. From the point of view of cohomology, the later means that one can integrate $\bar{\omega}_{WP}$ over in Fenchel-Nielsen coordinates and use the answer to study $\overline{\mathcal{M}}_g^C$.

Masur's formula established that the form $\bar{\omega}_{WP}$ has singularities along D. Wolpert deals with this by first showing that $\bar{\omega}_{WP}$ is a closed, positive $(1,1)$ current and then that $\bar{\omega}_{WP}$ is the limit of smooth, closed, positive $(1,1)$ forms.

5.2.2 How Rational is $\frac{1}{\pi^2}\bar{\omega}_{WP}$?

The second part of the argument is to establish the rationality of $\frac{1}{\pi^2}\bar{\omega}_{WP}$. According to Harer's results outlined in Chapter 3, section 3.3 and in Chapter 4 as well, $H_2\left(\mathcal{M}_g; \mathbb{Q}\right)$ is rank 1 when $g \geq 3$ [9]. Using this in addition to the Mayer-Vietoris spectral sequence gives $H_2\left(\overline{\mathcal{M}}_g; \mathbb{Q}\right) \cong H^{6g-8}\left(\overline{\mathcal{M}}_g; \mathbb{Q}\right)$, which has rank $2 + [g/2]$. We now proceed to describe dual bases for H_2 and H_{6g-8}.

The divisor $D = \overline{\mathcal{M}}_g - \mathcal{M}_g$ is made up of $1 + [g/2]$ components $D_0, \cdots, D_{[g}$ where D_i consists of generic surfaces with a node which are obtained by collapsing a curve to a point. For D_0 the curve is non-separating, while for

D_i, $i > 0$, the curves separate the surface into pieces of genus i and $g - i$. The Poincaré dual of $\bar{\omega}_{WP}$ and $D_0, \cdots D_{[g/2]}$ gives $2 + [g/2]$ classes in $H_{6g-8}\left(\overline{M}_g\right)$. The following Hermitian product is a direct consequence:

$$< \phi, \psi >_T = \int_X \phi\bar{\psi}K_1 + \phi\psi K_2.$$

Note that K_1 must be a $(-1, -1)$ tensor and K_2 a $(-3, 1)$ tensor; with respect to $<, >_T$, both must be $\mathrm{PSL}(2, \mathbb{R})$ invariant. This constraint gives only the possibility that K_1 is a multiple of the hyperbolic area element, and K_2 is zero. Hence, $<, >_T$ is a multiple of $<, >_{WP}$!

5.2.3   The Complex Structure on the Teichmüller Space

Wolpert [3] uses equation (5.2) to investigate the complex structure on $\mathcal{T}_g^s$. The investigation in question involves the Lie bracket formula, the duality formula and the formula

$$t_{\alpha_1} t_{\alpha_2} l_\beta + t_{\alpha_2} t_\beta - \alpha_1 + t_\beta t_{\alpha_1} l_{\alpha_2} = 0.$$

By skew-symmetry $t_{\alpha_1} l_{\alpha_2} = -t_{\alpha_2} l_{\alpha_1}$; consequently, $t_{\alpha_1} t_{\alpha_2} l_\beta [t_{\alpha_2}, t_\beta] l_{\alpha_1} = 0$. Dividing by l_β and taking β in a sequence which is uniformly distributed gives

$$< t_{\alpha_1}, t_{\alpha_2} > + 3\pi(g - 1)\omega_{WP}\left(\lim_j (1/l_{\beta_j})[t_{\alpha_2}, t_{\beta_j}], t_{\alpha_1}\right) = 0.$$

Since α_1 is arbitrary,

$$t_\alpha + 3\pi(g - 1)J\lim_j (1/l_{\beta_j})[t_\alpha, t_{\beta_j}] = 0.$$

Wolpert then derived the formula for the complex structure on the Teichmüller space $\mathcal{T}_g^s$:

$$Jt_C = 3\pi(g - 1)\lim_j \left(1/l_{\beta_j}\right)\left[t_C, t_{\beta_j}\right]. \qquad (5.11)$$

5.3   Projectivity of the Deligne-Mumford Moduli Space

As a final illustration of the importance of the Weil-Petersson geometry, we will give a sketch of Wolpert's proof that the Deligne-Mumford moduli space

$\overline{\mathcal{M}}_g$ is projective. The outline is relatively simple. First, one has to show that the Weil-Petersson metric ω_{WP} extends to $\bar{\omega}_{WP}$ on $\overline{\mathcal{M}}_g$. Next, using the fact that $\frac{1}{\pi^2} \bar{\omega}_{WP}$ is rational, it follows that $\frac{n}{\pi^2} \bar{\omega}_{WP}$ is the first Chern class c_1 of a line bundle L for some positive integer n. Now since ω_{WP} is Kähler, we deduce that L is positive. A theorem of Kodaira provides the imbedding $\overline{\mathcal{M}}_g \to \mathbb{C}P^n$. Technical subtleties do arise which make things a bit complicated. For instance, the extension ω_{WP} is smooth in Fenchel-Nielsen coordinates but otherwise difficult to evaluate in other systems of coordinates. Secondly, surfaces with several nodes lie on the intersection of several divisors. The divisors define cycles, and thus classes in $H_{6g-8}(\overline{\mathcal{M}}_g; \mathbb{Q})$ or, by Poincaré duality, classes in $H^2(\overline{\mathcal{M}}_g; \mathbb{Q})$. Recall that $H^2(\overline{\mathcal{M}}_g; \mathbb{Q})$ has rank $2 + [g/2]$ (valid only for $g \geq 3$). In actual fact, these are only $1 + [g/2]$ classes of divisors. Obviously, we need another class. Set $\epsilon \sim \mathcal{M}_1^1$. Computing the intersection matrice between the divisors, one finds that it is nonsingular. The next step is to evaluate the integrals $\int_\epsilon \bar{\omega}_{WP}$ and $\int_{\epsilon'} \bar{\omega}_{WP}$, where ϵ and ϵ' both describe an analytic 2 cycle in the the divisor D. Wolpert computes directly that

$$\int_{\overline{\mathcal{M}}_1^1} \bar{\omega}_{WP} = \frac{\pi^2}{6}.$$

Passing to another class (a 4-fold cover to be more accurate) gives

$$\int_{\overline{\mathcal{M}}_0^4} \bar{\omega}_{WP} = \frac{2\pi^2}{3}$$

and since Σ_1^4 double covers Σ_0^4,

$$\int_{\overline{\mathcal{M}}_0^4} \bar{\omega}_{WP} = \frac{\pi^2}{3}.$$

Using equation (5.1) allows us to show that $\bar{\omega}_{WP}$ restricts, that is to say

$$\begin{aligned}
\bar{\omega}_{wp|\epsilon'} &= \bar{\omega}_{WP} \quad \text{on } \overline{\mathcal{M}}_1^1 \\
\bar{\omega}_{wp|\epsilon'} &= \bar{\omega}_{WP} \quad \text{on } \overline{\mathcal{M}}_0^4.
\end{aligned}$$

This proves not only that the collections are bases, but that they further simultaneously evaluate $\bar{\omega}_{WP}$ and also show that $\frac{1}{\pi^2} \bar{\omega}_{WP}$ is indeed rational.

Let us pause a bit to look at our earlier assertion that $\frac{n}{\pi^2} \bar{\omega}_{WP}$ is the first Chern class $c_1(L)$ on the Deligne-Mumford moduli space. Observe that

$\frac{1}{\pi^2} \bar{\omega}_{WP}$ is a rational, closed positive $(1,1)$ current on $\mathcal{M}_g$. The general definition for $c_1(L)$ is

$$c_1(L) = \frac{1}{2\pi i} \log \partial \bar{\partial} \|s\|^2;$$

where by s we mean any section of the line bundle L, and $\|\ \|$ is a metric for L. Heuristically, if

$$\|s\|^2 e^{1/\log\left(\frac{1}{|z|^2}\right)}, \ |z| < 1,$$

then we have

$$\partial \bar{\partial} \log \|s\|^2 = \partial \bar{\partial} \frac{1}{\log(1/|z|^2)} = \frac{1}{|z|^2 \left(\log(1/|z|^2)\right)^3},$$

the principal term of Masur's formula. From our calculation, $e^{1/\log(1/|z|^2)}$ is a $\mathbb{C}^\infty$ metric of positive curvature in the sense of currents. Similarly, Wolpert finds that $\bar{\omega}_{WP}$ is the positive curvature form of a $\mathbb{C}^\infty$ metric for a line bundle over $\overline{\mathcal{M}}_g$. An argument of Richberg is then used to replace the metric in the line bundle by a smooth metric with positive curvature form. The statement then follows.

5.4 References

[1] Alvarez, Orlando: *Differential Geometry in String Models*, in **Unified String Theories**, Eds. M. Green and D. Gross (World Scientific, 1985) 103-139.

[2] Wolpert, S. : *The Fenchel-Nielsen Deformations* Ann. of Math. 115 (1982) 501-528.

[3] Wolpert, S. : *On the Symplectic Geometry of Deformations of a Hyperbolic Surface*, Ann. of Math. 117 (1983) 207-234.

[4] Wolpert, S. : *Thurston's Riemannian Metric for Teichmüller Spaces* Jour. Diff. Geometry 23 (1986) 143-174.

[5] Wolpert, S. : *On the Weil-Petersson Geometry of the Moduli Space of Curves*, Amer. Jour. Math. 107 (1985) 969-998.

[6] Masur, H. : *The Extension of the Weil-Petersson Metric to the Boundary of Teichmüller Space*, Duke Math. Journal 43 (1976) 623-635.

[7] Gardiner, F. P. : *Schiffer's Interior Variation and Quasiconformal Mapping*, Duke Math. Journal 42 (1975) 371-380.

[8] Goldman, W. : *The Symplectic Nature of Fundamental Groups of Surfaces*, Adv. in Math. 54 (1984) 200-225.

[9] Harer, J. : *The Second Homology of the Mapping Class Group of an Orientable Surface*, Inv. Math. 72 (1982) 221-239.

Chapter 6

Gauge Theories on Riemann Surfaces I: Bundles

Gauge theories are equations for connections which are invariant under gauge transformations. The gauge theoretical point of view has led to some spectacular developments in the geometry, topology and physics of three and four dimensions, but has also generated new insights into two-dimensional geometry, the geometry of Riemann surfaces. A Riemann surface is a connected, one-dimensional complex manifold. A Riemann surface can also be regarded as a real, two-dimensional oriented differentiable manifold.

Gauge theories have elicited a deep interaction among mathematicians and physicists. The initial step was taken in 1954 by Yang and Mills [1], when they introduced the concept of non-abelian gauge theory as a generalization of Maxwell's theory of Electromagnetism. The Yang-Mills theory involves a self-interaction among gauge fields. Around the same time, mathematicians were putting a final touch on their theory of fiber bundles [2], but this remarkable achievement was largely unknown to the physics community. It would take about ten years to recognize that Yang-Mills theory in fact, corresponds to the geometrical description of fiber bundles. Lubkin [3], Hermann [4] and Trautman [5] were credited with making this connection, although at that time very few of the implications of their work were explored. Wu and Yang [6] pointed out to the physics community the potential use of differential geometric methods of fiber bundles in gauge theories.

They showed, for instance, how the long-standing problem of the Dirac string for the magnetic monopole [7] could be solved using overlapping coordinate patches with gauge potentials differing by a gauge transformation; for mathematicians, the necessity of using coordinate patches is a trivial consequence of the fact that non-trivial fiber bundles cannot be described by a single gauge potential over the entire coordinate space. That same year, Belavin, Polyakov, Schwarz and Tyupkin [8] discovered a remarkable finite action solution of the Euclidean SU(2) Yang-Mills gauge theory, today known as the instanton. The instanton has self-dual or anti-dual field strength and carries a non-vanishing topological quantum number. This number, it turns out, is the integral of the second Chern class Chern class, an integer characterizing the topology of an SU(2) principal fiber bundle.

't Hooft [9] later showed that the instanton provides a mechanism for breaking chiral $U(1)$ symmetry and solving the long-standing problem of the ninth axial current, together with a possible mechanism for the CP-violation and anomalies arising from the conservation of the fermion number.

The instanton revealed the existence of a periodic structure of the Yang-Mills vacua. Jackiw and Rebbi [10], and Callan-Dashen-Gross [11] established in 1976 that the instanton action gives the lowest order approximation to the quantum mechanical tunneling amplitude between these states. The true ground state of the theory becomes the coherent mixture of all states. We begin our exploration of fiber bundles with a focus on their geometry.

6.1   Geometry of Fiber Bundles

We begin with a definition.

Definition 6.1 *Definition (Fiber Bundles)*

Consider M, a manifold which we shall refer to as the base manifold. Let M', another manifold, be the fiber. A fiber bundle E over M with fiber M' is a manifold which is locally a direct product of M and M'. In other words, if M is covered by a set of local coordinate neighborhoods $\{U_i\}$, then the bundle E is topologically described in each neighborhood U_i by the product manifold

$$S = U_i \times M'. \tag{6.1}$$

We point out that formula (6.1) does not provide sufficient information about the global topology of E since it is essentially a local product. Therefore, in order to determine the global topology of E, we introduce at this point a set of transition functions $\{\Phi_{ij}\}$. These are essential in telling us how the fiber manifolds match up in the overlap between two neighborhoods, say $U_i \cap U_j$. The explicit form for the transition function is

$$\Phi_{ij} : M'_{|U_i} \to M'_{|U_j} \tag{6.2}$$

in $U_i \cap U_j$. We learn from (6.2) that although the local topology of a given bundle is trivial, the global topology as determined by the transition functions may be quite complicated due to the relative twisting of neighboring fibers. It is precisely because of this fact that fiber bundles are called twisted products in the mathematical literature.

6.1.1   Non-Trivial Fiber Bundle: The Möbius Strip

The Möbius strip is constructed as follows. As a base manifold M, we take a circle S^1 parametrized by the angle θ. Let us cover S^1 by two semi-circular neighborhoods $U_\pm$:

$$U_+ = \{\theta : -\epsilon < \theta < \pi + \epsilon\};$$

$$U_- = \{\theta : \pi - \epsilon < \theta < 2\pi + \epsilon = 0 + \epsilon\}.$$

Next, we take the fiber M' to be an interval in the real line with coordinates $t \in [-1, 1]$, the main advantage of this procedure being to express the bundle in terms of two local pieces: $U_+ \times M'$ with coordinates (θ, t_+) and $U_- \times M'$ with coordinates (θ, t_-), and the transition relating t_+ to t_- in $U_+ \cap U_-$. The transition functions are

$$t_+ = t_- \text{ in the region } A = \{\theta : -\epsilon < \theta < \epsilon\}$$

$$t_+ = t_- \text{ in the region } B = \{\theta : \pi - \epsilon < \theta < \pi + \epsilon\}.$$

The process of identifying t with t_- in region B twists the bundle and gives it the non-trivial global topology of the Möbius strip.

6.1.2   Trivial Bundles

When all the transition functions reduce to the identity, the global topology of the bundle is that of the direct product:

$$E = M \times M'. \tag{6.3}$$

Such bundles are referred to as trivial fiber bundles, or more popularly, as trivial bundles. Here is an example of a trivial bundle. If we had set $t_+ = t_-$ in region A and B above, we would have found a trivial bundle equal to the cylinder $S^1 \times [-1, 1]$. A powerful theorem classifies fiber bundles as follows: any fiber bundle over a contractible base is trivial. Non-trivial fiber bundles can only be constructed when the global topology of the base space is non-trivial.

6.1.3   Sections of Fiber Bundles

A cross section or simply a section s of a fiber bundle E consists of assigning a preferred point $s(x)$ on each fiber to each point x of the base manifold M. A local section is a section which is only defined over a subset of M. Sections can be thought of as functions from U_i into M'. Note that the existence of global sections depends on the global geometry of the bundle E. Not all fiber bundles have global sections.

6.1.4   Pullback Bundles

Consider a fiber bundle E over the base manifold M with fiber M'. Let $M'' \to M$ denote the map obtained by h^*E; it is defined by copying the fiber of E over each point $x = h(x')$ in M x' in M''. We write $M'' \times E$ for the pairs (x', e); this gives the following explicit form

$$E' = h^*E = \{(x', e) \in M'' \times E\}, \tag{6.4}$$

such that $\pi(e) = h(x')$. From equation (6.4), we deduce that E' is a subset of $M'' \times E$; it is actually obtained by restricting oneself to $\pi(e) = h(x')$.

   To extract the transition functions of the pullback bundle E', we consider $\{U_i\}$, the covering of M for which E is locally trivial over U_i. The next step is to prove that $E' = h^\star E$ is also locally trivial. This can be done using the covering of M'' given by $\{h^{-1}|U_i\}$. The correponding transition functions of the pullback bundle are:

$$\Phi'_{ij} = (h' \Phi_{ij})(x') = \Phi_{ij}(h(x')). \tag{6.5}$$

6.2 Vector Bundles

We begin by collecting the necessary ingredients. Let E be a bundle with a k-dimensional real fiber $M = \mathbb{R}^k$ over an n-dimensional base space M; here, k stands for the bundle dimension, i.e. $\dim(E) = k$ (in reality, this is the dimension of the fiber alone, the total dimension of E being $(n+k)$.) E is called a vector bundle if its transition functions belong to the group $GL(k; \mathbb{R})$ rather than to the full group of diffeomorphisms (that is, to the group of differentiable transformations which are in one-to-one and invertible) of $\mathbb{R}^k$. A nice feature of the group $GL(k; \mathbb{R})$ is that it preserves the usual operations of addition and scalar multiplication on a vector space; consequently, the fibers of E inherit the structure of a vector space. A vector bundle can be thought of as a family of vector spaces (the fibers) which are parametrized by the base space M. This notion extends to the case of complex vector bundles if we replace $\mathbb{R}^k$ by $\mathbb{C}^k$ and $GL(k; \mathbb{R})$ by $GL(k; \mathbb{C})$. Listed below are some useful properties of vector bundles.

The origin $\{0\}$ of $\mathbb{C}^k$ or $\mathbb{R}^k$ is preserved by the general linear group and represents an element of the fiber of a vector bundle, the zero-section. Vector bundles have many global sections, but few have global sections which are non-zero everywhere.

The vector space structure on the fibers of a given vector bundle allows the definition of pointwise addition or scalar multiplication of sections. Let $s(x)$ and $s'(x)$ denote two local sections of E. One can define the local section $(s + s')(x) = s(x) + s(x')$ by adding the values in the fibers. This amounts to defining a smooth function $f(x)$ on M, which is then used in writing down the derived new section $[fs](x) = s(x)f(x)$ by pointwise scalar multiplication in the fibers.

6.2.1   Line Bundles

Definition 6.2 *Definition A line bundle is a vector bundle with a one-dimensional vector space as its fiber. It is a family of lines parametrized by the base space M. The Möbius strip discussed in Subsection 6.1.1 provided us with a good illustration of line bundles. Suppose for instance, that we replace the interval $[-1, 1]$ by the real line $\mathbb{R}$ in our original Möbius strip example. Hence we have a non-trivial real line bundle over the circle. Replacing now $[-1, 1]$ by the complex numbers $\mathbb{C}$ gives a line bundle isomorphic to $S^1 \times \mathbb{C}$. The message here is that the nature of the line bundles depends very much on whether we write the transition functions on the left or on the right.*

6.2.2   Tangent and Cotangent Bundles

Let $T(M)$ and $T^\star(M)$, respectively denote the tangent bundle and the cotangent bundle. These are real vector bundles whose fibers at a point $x \in M$ are given by the tangent space $T_x(M)$ or the cotangent space $T_x^\star(M)$. The local frame for the tangent bundle $T(M)$ reads

$$\{\partial/\partial x_1, \cdots, \partial/\partial x_n\},$$

and similarly, for the cotangent bundle

$$\{dx_1, \cdots, dx_n\};$$

note that $x = (x_1, \cdots, x_n)$ is a local coordinate system defined in some neighborhood $U \in M$.

If U' is another neighborhood in M with local coordinates x', then we can express the sought transition functions in $U \cap U'$ as

$$\frac{\partial}{\partial x_i} = \frac{\partial}{\partial x'_j} \cdot \frac{\partial x'_j}{\partial x_i} \quad \text{on } T(M)$$

$$dx_i = dx'_j \cdot \frac{\partial x_i}{\partial x_j} \quad \text{on } T^\star(M).$$

The complexified tangent and cotangent bundles $T(M) \otimes \mathbb{C}$ and $T^\star(M) \otimes \mathbb{C}$ of a real manifold M are defined by allowing the coefficients of the frame $\{\frac{\partial}{\partial x_i}\}$

and $\{dx_i\}$ to be complex. In this case, the complex tangent bunndle $T_{\mathbb{C}}$ is the sub-bundle of $T(M) \otimes \mathbb{C}$ spanned by the holomorphic tangent vectors $\partial/\partial z_j$, where z_j denotes the local complex coordinates of the complex manifold M. The complex dimension of $T_{\mathbb{C}}(M)$ is half the real dimension of $T(M)$.

6.2.3 Building Fiber Bundles

To a vector space V, we associate its dual space V^*, a set of linear functionals. If V and W are a pair of vector spaces, we can define the Whitney sum $V \oplus W$ and the tensor product $V \otimes W$. These and other constructions can be carried over the vector bundle case, as described below.

Let us first recall some important facts about the dual space V^*, the space of linear functionals. An element $v \in V^*$ is a linear map $v^* : V \to \mathbb{R}$. Since the sum and scalar multiple of linear maps are again linear map V^* is thus a vector space. Choosing $\{e_1, \cdots, e_k\}$ as a basis for V and $v^* \in V^*$ gives $v^*(e_j z^j) = z^j v^*(e_j)$. This means that the action of v^* on a given section is determined by the value of the linear map on the basis. The dual basis $\{e^{*1}, \cdots, e^{*k}\}$ of the dual space V^* is defined by the linear functional

$$e^{*i}(e_j) = \delta^i_j \;\Rightarrow\; e^{*i}\left(e_j z^j\right) = z^i.$$

The e^{*i} define coordinates on V; similarly, the e_j define coordinates on V^*. This is so because of the relation

$$\dim(V) = \dim(V^*) = k.$$

To obtain a new dual basis, one changes bases and set $e_i = e_j \Phi_{ji}$. The new basis reads:

$$\sum e^{*i} = \Phi_{ij}^{-1} e^{*ij} e^{*ij} \left(\Phi^t\right)^{-1}_{ji}. \tag{6.6}$$

We learn from (6.6) that the dual basis transforms just as a set of coordinates on V would transform.

Dual vector spaces arise naturally whenever we have two vector spaces V and W together with a non-singular inner product $(v, w) \in \mathbb{R}$ or $\mathbb{C}$, where $v \in V$ and $w \in W$. Because (v, w) is a linear functional on V, we can regard w as an element of the dual space V^* whose action is

$$w(v) = (v, w).$$

Owing to the fact that the inner product is non-singular, one may identify W with V. Conversly, V and V^* each possess a natural inner product defined by the action of v^* on v:

$$(v, v^*) = v^*(v).$$

We may regard V itself as a space of linear functionals dual to V^* if the action of $v \in V$ is

$$v(v^*) = (v, v^*) = v^*(v).$$

When V is finite dimensional, $V^* = V$, whereas, for infinite dimensional V, this property fails.

We now focus on the Whitney sum bundle. The Whitney sum $V \oplus W$ of two vector spaces V and W is defined to be the set of all pairs (v, w). The vector space structure of (v, w) is:

$$(v, w) + (v', w') = (v + v', w + w')$$

and

$$\lambda(v, w) = (\lambda v, \lambda w).$$

To show that V and W are subspaces of $V \oplus W$, suffices to identify v with $(v, 0)$ and w with $(0, w)$. Let $\{e_i\}$ and $\{f_j\}$ denote bases for V and W respectively. It then follows that $\{e_i, f_i\}$ is a basis for $V \oplus W$, and this implies

$$\dim(V \oplus W) = \dim(V) + \dim(W).$$

Now let us consider E and F, two vectors bundles over M. The fiber of the Whitney sum bundle $E \oplus F$ is obtained by taking the Whitney sum of the fibers E and F at each point $x \in M$. When dim $(E) = j$ and $\dim(F) = k$, the transition functions of E and M' are just the $j \times j$ matrices Φ and the $k \times k$ matrices Ψ, respectively. Thus, the transition matrices of $E \oplus F$ are just the $(j + k) \times (j + k)$ matrices $\Phi \oplus \Psi$; they are given by the expression:

$$\begin{pmatrix} \Phi & 0 \\ 0 & \Psi \end{pmatrix} = \Phi \oplus \Psi.$$

The tensor product bundle $E \otimes F$ of E and F is obtained by taking the tensor product of the fibers of E and F at each point $x \in M$. The transition matrices for $E \otimes F$ are obtained by taking the tensor product of the transition functions of E and that of F.

6.2.4   Line Bundles Over Projective Spaces

We take $P_n(\mathbb{C})$ to be the set of lines through the origin in $\mathbb{C}^{n+1}$. Let $I^{n+1} = P_n(\mathbb{C}) \times \mathbb{C}^{n+1}$ be the trivial bundle of dimension $n+1$ over $P_n(\mathbb{C})$. Take L to be the sub-bundle of I^{n+1} defined by

$$L = \{(p, z) \in I^{n+1} \mid z \in p\}. \tag{6.7}$$

The fiber of L over a point p of $P_n(\mathbb{C})$ is just the set of points in $\mathbb{C}^{n+1}$ which belong to the line p. A section s_j of L can be written as

$$s_j(p) = (\zeta_k^{(j)})^{-1} s_j(p),$$

where $\zeta_k^j = z_i/z_j$ are coordinates. It is a fact that the transition functions are holomorphic; consequently, L inherits the holomorphicity and becomes a holomorphic line bundle. As for the dual bundle L', it has sections $s_j^\star$ satisfying $s_j^\star(s_j) = 1$. The transition function is

$$s_k^\star = s_j^\star \, \zeta_j^{(j)}.$$

Owing to the fact that $s_j^\star(p) = z_j$, we regard the $\{s_j^\star\}$ as homogeneous coordinates on $P_n(\mathbb{C})$. The s_j are actually meromorphic sections of L. We write L^k as

$$\begin{array}{ll} L^\star \otimes \cdots \otimes L^\star & \text{if } k < 0 \\ L^0 = I & \text{trivial line bundle} \\ L \otimes \cdots \otimes L & k > 0 \\ L \otimes L^\star = I, & L^j \otimes L^k = L^{j+k}. \end{array} \tag{6.8}$$

Any line bundle over $P_n(\mathbb{C})$ is isomorphic to L^k for some uniquely defined integer k. This integer is related to the first Chern class of L^k, as will become clear later on. For now however, let us consider $T_c(P_n(\mathbb{C}))$ and $T_c^\star(P_n(\mathbb{C})) = \Lambda^{1,0}(P_n(\mathbb{C}))$, respectively, the complex tangent and cotangent spaces. We have:

$$\begin{aligned} I \oplus T_c(P_n(\mathbb{C})) &= L^\star \oplus \cdots \oplus L^\star \\ I \oplus T_c^\star(P_n(\mathbb{C})) &= L \oplus \cdots \oplus L. \end{aligned}$$

Although this identity does not preserve the holomorphic structure, it is clearly an isomorphism between complex vector bundles.

6.2.5   Principal Bundles

The presentation above makes it crystal clear that a vector bundle is a fiber bundle whose fiber M is a linear vector space, and whose transition functions belong to the general linear group of M. By contrast, a principal bundle P is a fiber bundle whose fiber is actually a Lie group G. The transition functions of P belong to G and act on G by left multiplication. A right action of G on P is the result of exploiting the commutation between right and left multiplication. Related to a principal bundle P are the frame bundle of a vector bundle E and the associated principal bundle. They can be constructed in the following outlined procedure. The fiber G_X of P at X is the set of all frames of the vector space M_X, which is the fiber of E over the point X.

With this in hand, we now consider a complex vector space $M = \mathbb{C}^k$ of dimension k. By definition, the fiber G of the frame bundle P is the collection of $k \times k$ non-singular matrices which form the group $GL(k; \mathbb{C})$, where G is the structure group of the vector bundle E. The associated principal bundle has the same transition functions as the vector bundle E. These transition functions are $\mathrm{GL}(k; \mathbb{C})$ group elements whose action on the fiber G are essentially through left multiplication. Hence with P, a principal G bundle and p, a representation of G on a finite-dimensional vector space V, one can define the associated vector bundle $P \times_p V$ by the equivalence relation on $P \times V$:

$$(p, p(g) \cdot v) \simeq (p \cdot g, v) \text{ for all } p \in P, v \in V, g \in G.$$

6.2.6   Yang-Mills Instanton as Principal Bundle

This is, to date, the most interesting and celebrated example of a principal bundle. The base of this instanton bundle is the compactified Euclidean space-time, i.e. the four-sphere S^4, and its fiber is the group SU(2), which we may recall is equivalent to S^3. We endow the base S^4 with coordinates (θ, ϕ, ψ, r). Similarly, for the fiber we have the coordinates (α, β, γ). The next step is to split S^4 into two pieces, D_+ and D_- whose boundaries are S^3. The intersection of D_+ with D_- along the equator of S^4 is parametrized by the Euler angles (θ, ϕ, ψ) of S^3. The representation $h(\theta, \phi, \psi)$ of SU(2) is

$$h = \frac{t - i\lambda \cdot x}{r},$$

$$\begin{cases} x + iy &= r\cos\frac{\theta}{2}\, e^{(i/2)(\psi+\phi)} \\ z + it &= r\sin\frac{\theta}{2}\, e^{(i/2)(\psi-\phi)} \end{cases}.$$

The fiber coordinates are given in a somewhat similar fashion by SU(2) matrices $g(\alpha, \beta, \gamma)$. From there, we have a good picture of what the local bundle patches ought to look like. Explicitly, they are

$$D_+ \times \mathrm{SU}(2), \text{ with coordinates } (\theta, \phi, r; \alpha_+, \beta_+, \gamma_+)$$
$$D_- \times \mathrm{SU}(2), \text{ with coordinates } (\theta, \phi, \psi, r; \alpha_-, \beta_-, \gamma_-).$$

In the intersection area $D_+ \cap D_-$, transitions from the SU(2) fiber $g(\alpha_+, \beta_+, \gamma_+)$ to $g(\alpha_-, \beta_-, \gamma_-)$ can be built using multiplication by the SU(2) matrix $h(\theta, \phi, \psi)$. The result is

$$g(\alpha_-, \beta_-, \gamma_-) = h^k(\theta, \phi, \psi)\, g(\alpha_+, \beta_+, \gamma_+). \tag{6.9}$$

When $k = 1$ we have the Hopf fibering of S^7 [2, 5], namely:

$$P = S^7.$$

This precisely fits the description of the bundle prescribed by the single-instanton solution of Belavin et al. [8]. More general instanton solutions describe bundles with different values of k.

6.2.7 Dirac's Magnetic Monopole

Dirac's magnetic monopole is in essence a principal $U(1)$ bundle over S^2. To explicitly construct such a bundle, we start with a base $M = S^2$ and fiber $M' = U(1) = S^1$. The $U(1)$ coordinate is labelled by $e^{i\psi}$, while those of the base S^2 are (θ, ϕ), $0 \le \theta < \pi$, $0 \le \phi < 2\pi$. As with the Yang-Mills instanton, we break S^2 into two hemispherical neighborhoods, D_- and D_+; we also take $D_+ \cap D_-$ to denote the boundary of the equatorial area parametrized by the angle ϕ. The local form of the bundle is therefore

$$D_+ \times U(1) \quad \text{with coordinates } (\theta, \phi;\, e^{i\psi_+})$$
$$D_- \times U(1) \quad \text{with coordinates } (\theta, \phi;\, e^{i\psi_-}).$$

To obtain a principal bundle out of this construction, one has to restrict the transition functions in such a way that they define only $U(1)$ functions of ϕ along the intersection $D_+ \cap D_-$. This means literally that one needs to relate the D_+ and D_- fiber coordinates. This is done by

$$e^{i\psi_-} = e^{in\phi_-} e^{i\phi_+}. \tag{6.10}$$

Note that n in equation (6.10) must be an integer for the resulting structure to be a manifold. The fibers ought to fit together perfectly when we complete a full revolution around the equator in ϕ. This is the topological version of the Dirac monopole quantization condition. Let us take a look at various values of n. For $n = 0$, we have the trivial bundle

$$P = S^2 \times S^1.$$

For $n = 1$,

$$P = S^3,$$

and we recover the Hopf fibering [2, 5]. This bundle describes a single charged Dirac monopole. Larger values of n correspond to more complicated monopole bundles.

6.3   References

[1] Yang, C. N. and Mills, R. L.: *Conservation of Isotopic Gauge Invariance*, Phys. Rev. 96 (1954) 191-195.

[2] Steenrood, N. : **The Topology of Fiber Bundles**, Princeton University Press, 1951.

[3] Lubkin, E.: *Geometric Definition of Gauge Invariance*, Ann. Phys. 23 (1963) 233-283.

[4] Hermann, R.: **Vector Bundles in Mathematical Physics**, Vols. I and II, 1970, W. A. Benjamin, Inc.

[5] Trautman, A.: *Fiber Bundles Associated With Space-Time*, Reports on Math. Physics 1 (1970) 29-62.

[6] Wu, T. T. and Yang, C. N.: *Concept of Nointegrable Phase Factors and Global Formulation of Gauge Fields*, Phys. Rev. D12 (1975) 3845-3857.

[7] Dirac, P. A. M.: *Quantized Singularities in the Electromagnetic Field*, Proc. Roy. Soc. London A133 (1931) 60-72.

[8] Belavin, A., Polyakov, A. M., Schwarz, A. S. and Tyupkin, Y. S.: *Pseudoparticle Solutions of the Yang-Mills Equations*, Phys. Lett. B59 (1975) 85-87.

[9] 't Hooft, G.: *Symmetry Breaking Through the Bell-Jackiw Anomalies*, Phys. Rev. Lett. 37 (1976) 8-11.

[10] Jackiw, R. and Rebbi, C.: *Vacuum Periodicity in a Yang-Mills Quantum Theory*, Phys. Rev. Lett. B37 (476) 172-175.

[11] Callan, C. G., Dashen, R. and Gross, D. J.: *The Structure of the Gauge Theory Vacuum*, Phys. Lett. B63 (1976) 334-340.

Chapter 7

Gauge Theories on Riemann Surfaces II: Connections

The physical potential of the theory of fiber bundles lies in connections on bundles. The connection 1-form, for instance, is well known to physicists as a gauge potential; similarly, the Yang-Mills field strength is defined as the curvature associated with the connection. Nowhere is this potential more pronounced than with Chern-Simons-Witten (CSW) theories: in the Hamiltonian approach of CSW theories, an appropiate description of the 3-manifold invariant (see Chapter 1, Section 4) requires the geometric quantization of the space of flat connections on a given Riemann surface [1, 2, 3, 4, 5, 6]. Chapter 8 is devoted to the quantization of CSW theories. For now, however, we aim to focus on the basics of connections on fiber bundles. Apart from its central role in gauge theories, the notion of connection plays an important role in the local differential geometry of fiber bundles. There, a connection defines a covariant derivative which contains a gauge field; connections are important in specifying the way in which a vector bundle E could be parallel-transported along a curve lying in the base manifold M, thus yielding information about holonomy and other related geometrical characteristics in the process. Below, we start our presentation with a description of connections on vector bundles. Connections on principal bundles are treated thereafter.

7.1   Connections on Vector Bundles

Physics spells out the need to differentiate sections of a vector bundle. For instance, a charged scalar field in Quantum Electrodynamics (QED) is regarded as a section of a complex line bundle associated with a $U(1)$ bundle, $P(M, U(1))$ [7]. The differentiation of sections ought to be done covariantly if there is to be a consistent theory with a gauge-invariant action. Lack of such invariance spoils the consistency of gauge theories and gives rise to pathologies called anomalies (Chapters 9, 10, 11, 12 and 13 are devoted to the study of various forms of anomalies in gauge theories in this book). The Levi-Civita connection on a surface in $\mathbb{R}^3$ illustrates this case perfectly. We consider the unit sphere $S^2 \subset \mathbb{R}^3$ as a specific working example. S^2 is parametrized by the coordinates

$$x(\theta, \phi) = (\sin\theta\cos\phi, \, \sin\phi\cos\theta, 0),$$

with $0 \leq \theta \leq \pi$ and $0 \leq \phi \leq 2\pi$. The resulting Riemannian metric

$$g_{ij} = \begin{pmatrix} \partial_\theta x \cdot \partial_\theta x & \partial_\theta x \cdot \partial_\phi x \\ \partial_\theta x \cdot \partial_\phi x & \partial_\phi x \cdot \partial_\phi x \end{pmatrix} = \begin{pmatrix} 1 & 0 \\ 0 & sin^2\theta \end{pmatrix}$$

implies

$$ds^2 = d\theta^2 + \sin^2\theta \, d\phi^2.$$

There are two vector fields, namely

$$a_1 = \partial_\theta x = (\cos\theta\cos\phi, \, \cos\theta\sin\phi, \, -\sin\theta)$$

and

$$a_2 = \partial_\phi x = (-\sin\theta\sin\phi, \, \sin\theta\cos\phi, 0),$$

which are tangent to the surface and span the tangent space. Derivatives can be decomposed into tangential components proportional to a_1 and a_2, and a normal component $\hat{n}$ proportional to x. Identifying a_1 and a_2 with the basis $\partial/\partial\theta$ and $\partial/\partial\phi$ for the tangent space follows mainly from the relations:

$$\frac{\partial f(x)}{\partial\theta} = a_1 \cdot \frac{\partial f}{\partial x}$$

$$\frac{\partial f(x)}{\partial\phi} = a_2 \cdot \frac{\partial f}{\partial x}$$

where $f(x)$ is a function on $\mathbb{R}^3$.

We now focus on the differentiation to which we referred earlier. We want to differentiate tangential vector fields with respect to the surface. As a first step, we write down the ordinary partial derivatives

$$\partial_\theta(a_1) = (-\sin\theta\cos\phi, -\sin\theta\sin\phi, -\cos\theta)$$

$$\partial_\phi(a_1) = \partial_\theta(a_2) = (-\cos\theta\sin\phi, \cos\theta\cos\phi, 0) = \frac{\cos\theta}{\sin\theta} a_2$$

$$\partial_\phi(a_2) = (-\sin\theta, \cos\phi, -\sin\theta\sin\phi, 0) = -\sin^2\theta x - \cos\theta\sin\theta a_2.$$

The next step centers on finding the appropriate intrinsic covariant differentiation ∇_x, which is defined with respect to a tangent vector x. To produce it, we use a rather simple method: one takes the ordinary derivatives and projects them back to the surface. Under this procedure, ∇_x is actually the directional derivative obtained by throwing away the normal component of the ordinary partial derivative. Explicitly:

$$\nabla_{a_1}(a_1) = 0$$
$$\nabla_{a_1}(a_2) = \nabla_{a_2}(a_1) = \cot\theta\, a_2$$
$$\nabla_{a_2}(a_2) = -\cos\theta\sin\theta\, a_1.$$

∇ is called the Levi-Civita connection on S^2. Identifying (a_1, a_2) with $(\partial/\partial\theta, \partial/\partial\phi)$ gives rise to the formula

$$\nabla_{\partial/\partial\theta} \equiv \nabla_{a_1},$$

and

$$\nabla_{\partial/\partial\phi} \equiv \nabla_{a_2}.$$

As for the Christoffel symbol, they are given by

$$\nabla_{a_i}(a_j) = a_k\, \Gamma^k{}_{ij};$$

or

$$\nabla_{\partial_i}(\partial_j) = \Gamma^k{}_{ij}\, \partial_k,$$

where $\partial_1 = \partial/\partial\theta$, $\partial_2 = \partial/\partial\phi$. According to previous formulas,

$$\Gamma^2{}_{12} = \Gamma^2{}_{21} = \cot\theta$$
$$\Gamma^1{}_{22} = -\cos\theta\sin\theta$$
$$\Gamma^k{}_{ij} = 0 \text{ otherwise.}$$

A few definitions are in order. A curve $x(t) \subset S^2$ is a geodesic if the acceleration $\ddot{x}$ has only components normal to the surface, i.e.

$$\nabla_{\dot{x}}(\dot{x}) = 0. \tag{7.1}$$

The Levi-Civita connection provides a rule for the parallel transport of vectors on a surface. As an illustration, consider $x(t)$ a curve in S^2, and let $s(t)$ be a vector field defined along the curve. S is parallel-transported along the curve if it satisfies the equation

$$\nabla_{\dot{x}}(s) = 0.$$

Let x be the geodesic triangle in S^2 connecting the points $(1,0,0)$, $(0,1,0)$ and $(0,0,1)$. x consists of three circles:

$$x(t) = \begin{cases} \cos t, \sin t, 0) & t \in [0, \pi/2] \\ 0, \sin t, -\cos t & t \in [\pi/2, \pi] \\ (-\sin(t), 0, -\cos(t)) & t \in [\pi, 3\pi/2]. \end{cases}$$

Write the initial tangent vector

$$s(0) = (0, \alpha, \beta)$$

at $(1,0,0)$. By parallel-transporting $s(0)$ along $x(t)$ using the Levi-Civita connection, we obtain

$$s(t) = \begin{cases} -\alpha \sin t, \alpha \cos t, \beta & t \in [0, \pi] \\ -\alpha, \beta \cos t \, \beta \sin t & t \in [\pi/2, \pi] \\ \alpha \cos t - \beta, -\alpha \sin t & t \in [\pi, 3\pi/2] \end{cases}.$$

Owing to the fact that $\partial s/\partial t$ is normal to the surface, it is rather straightforward to see that $s(t)$ is continuous at the endpoints $\pi/2$ and π and satisfies $\nabla_{\dot{x}}(s) = 0$. A parallel translation around the geodesic triangle x changes s from $s(0) = (0, \alpha, \beta)$ to $s(3\pi/2) = (0, -\beta, \alpha)$, which represents a rotation through $\pi/2$, where $\pi/2$ is the area of the spherical triangle.

Consider a rotation generated by a vector parallel-transported around a curve. The holonomy is, by definition, the process of assigning to each closed curve a linear transformation measuring such a rotation. Holonomy matrices form a group, the holonomy group.

Below, we provide some properties of the connection $\nabla_X(s)$ and the total covariant derivative ∇. The focus is firstly on the connection $\nabla_X(s)$. Its linearity is expressed by

$$\nabla_X(s + s') = \nabla_X(s) + \nabla_X(s').$$

The linearity in X is

$$\nabla_{X+X'}(s) = \nabla_X(s) + \nabla_{X'}(s).$$

The connection often behaves like a first order differential operator, as shown by the formula

$$\nabla_X(sf) = s \cdot X(f) + (\nabla_X(s))f.$$

The tensoriality in X follows from

$$\nabla_{fX}(s) = f\nabla_X(s).$$

Note that $f(x)$ denotes a scalar function, X stands for a vector field and $s(x)$ is a section of E.

The linearity of the total covariant derivative ∇ is given by the expression

$$\nabla(s + s') = \nabla(s) + \nabla(s').$$

This also behaves as a first order differential operator, as exhibited by the relation

$$\nabla(sf) = s \otimes df + \nabla(s)f.$$

On the basis of these facts, we can derive a relationship between these two differential operators, namely:

$$(1). \nabla(s) = \nabla_{\partial/\partial x^\mu} \otimes dx^\mu$$

$$(2). \nabla_X(s) = < \nabla(s), X >,$$

where $X \in \mathbb{C}^\infty(T(M))$ and $\nabla(s) \in \mathbb{C}^\infty(E \otimes T^*(M))$.

7.2   Curvatures

In the framework of connections, the curvature measures the extent to which parallel transport is path dependent. For instance, when the curvature is trivial, parallel transporting a given path in M results in the identity transformation. (There are some exceptions though, the most notable of which is a path with holes.) For curved manifolds, the story is different since one gets non-trivial results: a parallel-transport around a geodesic triangle on S^2 gives a rotation equal to the area of the spherical triangle.

In order for us to evaluate the curvature we need to use parallel transport. As an illustration, consider a local coordinate chart $(x_1, x_2, \cdots)$ along with a square path $x(t)$ with vertices. Write the explicit form of the vertices as

$$(0,0,0,\cdots), (0, \tau^{1/2}, 0, \cdots), (\tau^{1/2}, \tau^{1/2}, 0, \cdots), (\tau^{1/2}, 0, 0, \cdots);$$

the holonomy matrix $\tau_{ij}(\tau)$ is obtained by traversing the path with the vertices above. The curvature matrix in this plane is thus

$$g_{ij} = \frac{d}{d\tau} \tau_{ij}(\tau)|_{\tau=0}. \tag{7.2}$$

Now, let's move our focus to tangent spaces. Here, the curvature is the commutator of the components D_μ of the basis for the horizontal subspace of $T(E)$, that is:

$$[D_\mu, D_{n u}] = -g^i{}_{j\nu}\, z^i \frac{\partial}{\partial z^i}; \tag{7.3}$$

the $g^i{}_{j\mu\nu}$ can be expressed in terms of Christoffel symbols:

$$g^i{}_{j\mu\nu} = \partial_\mu \Gamma^i{}_{\mu j} + \Gamma^i{}_{\mu k} \Gamma^k{}_{\nu j} - \Gamma^i{}_{\nu k} \Gamma^k{}_{\mu j}.$$

In the cotangent space approach, the curvature takes on the appearance of a matrix-valued two-form,

$$\tau^i{}_j = d\Gamma^i{}_j + \Gamma^i{}_k \wedge \Gamma^k{}_j = \frac{1}{2} \tau^i{}_{j\mu\nu}\, dx^\mu \wedge dx^\nu. \tag{7.4}$$

Note that $\tau^i{}_j z^j = d\omega^i + \Gamma^i{}_j \wedge \omega^j$ is the covariant differential of the one-form $\omega^i \in T^*(E)$; although ω^i has dz^k components, they cancel out in $\tau^i{}_j$.

It follows from all of this that the curvature measure, in effect, is the extent to which covariant differentiation fails to commute.

We now focus on the curvature operator,

$$g\,(X,Y)\,(s) \;=\; \nabla_X \,\nabla_Y\,(s) \;-\; \nabla_Y \,\nabla_X\,(s) \;-\; \nabla_{[X,Y]}\,(s), \qquad (7.5)$$

where

$$g\left(\frac{\partial}{\partial x^\mu},\frac{\partial}{\partial x^\nu}\right)(e_i) \;=\; e_j\,g^j{}_{i\mu\nu}.$$

Catalogued below are the properties of the curvature operator, with X and Y as vector fields, $s(x)$ a section and $f(x)$ a scalar function.

• *Tensoriality*

$$g\,(fX,Y)\,(s) \;=\; g\,(X,fY)\,(s) \;=\; g\,(X,Y)\,(fs) \;=\; fg\,(X,Y)\,(s).$$

• *Antisymmetry*

$$g\,(X,Y)\,(s) \;=\; -g\,(Y,X)\,(s).$$

• *Multilinearity*

$$g\,(X+X',Y)\,(s) \;=\; g\,(X,Y)\,(s) \;+\; g\,(X',Y)\,(s).$$

The total curvature g is a matrix-valued 2-form given by

$$\begin{aligned}
g\,(s) \;&=\; \nabla^2\,(s) \;=\; \nabla\,(e_j \otimes \Gamma^j{}_i\,z^i\,e_j \otimes dz^i)\\
&=\; e_k \otimes \Gamma^k{}_j \wedge \Gamma^j{}_i\,z^i\\
&\quad +\, e_k \otimes \left(d\Gamma^k{}_i\,z^i - \Gamma^k{}_j \wedge dz^j\right)\\
&\quad +\, e_k \otimes \Gamma^k{}_j \wedge dz^j + 0\\
&=\; e_k \otimes g^k{}_i\,z^i.
\end{aligned} \qquad (7.6)$$

The matrix $g \;=\; \|g^i{}_j\|$ can be written as

$$g \;=\; \frac{1}{2}\,g\left(\frac{\partial}{\partial x^\mu},\frac{\partial}{\partial x^\nu}\right)dx^\mu \wedge dx^\nu$$

acting on sections s.

7.3 Torsions, Connections, and Tangent Bundles

The cotangent space formulation

$$\nabla (s) = e_i \otimes dz^i (x) + e_i \otimes \Gamma^i{}_j z^j (x) \tag{7.7}$$

of the vector bundle connection ∇ has the advantage that it is independent of the coordinate system $\{x^\mu\}$ on M. Furthermore, multiple covariant differentiation of an invariant one-form such as $\rho_\mu \, dx^\mu$ is also independent of the connection chosen on the cotangent bundle $T^*(M)$. In order to differentiate the tensor components $z^i{}_{j\mu}$ of the covariant derivative of a vector bundle section $s(x) = e_i z^i(x)$, one needs to specify a connection on $T^*(M)$. Torsion is, on the other hand, a property of the connection on the tangent bundle, a crucial ingredient for any study of double covariant derivatives.

Consider $\{\Gamma^i{}_{\mu j}\}$, the Christoffel symbol on the vector bundle E. Let $\{\gamma^\mu{}_{\mu\lambda}\}$ denote the Christoffel symbol on the tangent $T(M)$. We then have

$$
\begin{aligned}
z^i{}_{;\mu;\nu} &= \partial_\nu \left(\partial_\mu z^i + \Gamma^i{}_{\mu j} z^j \right) \\
&+ \Gamma^i{}_{\nu j} \left(\partial_\mu z^i + \Gamma^j{}_{\mu k} z^k \right) \\
&- \gamma^\lambda{}_{\mu\nu} \left(\partial_\lambda z^i + \Gamma^i{}_{\lambda j} z^j \right)
\end{aligned}
$$

for the double covariant derivative of a given section $s(x) = e_i z^i(x)$. The sign in front of $\gamma^\lambda{}_{\mu\nu}$, it should be noted, comes from the requirement of lowering indices in order to obtain the connection on $T^*(M)$. The use of the commutator of double covariant differentiation on a section gives rise to

$$z^i{}_{;\mu;\nu} - z^i{}_{;\nu;\mu} = -g^i{}_{j\mu\nu} z^j - T^{\lambda\nu} z^i. \tag{7.8}$$

We point out the presence of a new tensor in equation (7.8), i.e. the torsion

$$T^\lambda{}_{\mu\nu} = \gamma^\lambda{}_{\mu\nu} - \gamma^\lambda{}_{\nu\mu}.$$

By definition, we regard the torsion operator on $T(M)$ as

$$T(X,Y) = \nabla_X Y - \nabla_Y X - [X,Y],$$

that is, a vector field with components

$$T\left(\frac{\partial}{\partial x^\mu}, \frac{\partial}{\partial x^\nu} \right) = \left(\gamma^\lambda{}_{\mu\nu} - \gamma^\lambda{}_{\nu\mu} \right) \frac{\partial}{\partial x^\lambda}.$$

Once a metric of the type $(X, Y) = g_{\mu\nu}\, x^{mu}\, y^{\nu}$ has been chosen, the Levi-Civita connection on the tangent bundle $T(M)$ is uniquely defined because

$$T(X, Y) = 0, \text{ torsion-free property}$$
$$d(X, Y) = (\nabla X, Y) + (X, \nabla Y), \text{ covariant consistancy of metric.}$$

7.4   Connections on Various Bundles

Let E and E', respectively, denote two dual vector bundles with dual frame bases $\{e_i\}$, and $\{e'^i\}$. The connection ∇' on E' is defined by requiring that the natural inner product between sections s and s' be differentiated according to the following rule:

$$d < s, s' > = \langle \nabla (s), s' \rangle + \langle s, \nabla' (s') \rangle.$$

This implies that one can explicitly write $\nabla(e_i)$ and $\nabla'(e'^i)$ as follows:

$$\nabla(e_i) = e_j\, \Gamma^j{}_{\mu i}\, dx^\mu$$
$$\nabla'(e'^j) = e'^j\, \Gamma^i{}_{\mu j}\, dx^\mu.$$

Endowing E with a fiber metric allows one to identify E with E' using a conjugate linear isomorphism. The connection ∇ is said to be Riemannian if $\nabla = \nabla'$, that is,

$$\Gamma^i{}_{\mu j} = -\Gamma^j{}_{\mu i}, \tag{7.9}$$

relative to an orthonormal frame basis. The curvature of a Riemannian connection relative to an orthonormal basis is antisymmetric:

$$g^i{}_j = -g^j{}_i. \tag{7.10}$$

The Levi-Civita connection on $T(M)$ is the unique torsion-free Riemannian connection.

The question of describing the torsion in terms of the Whitney sum bundle then arises. In order to work it out, let us recall what a Whitney sum bundle $E \oplus F$ is; it is obtained by taking the Whitney sum of the fibers of E and F at each point $x \in M$. With these descriptions in hand, we proceed to endow E and F, two vector bundles, with connections ∇ and ∇' respectively. The resulting natural connection $\nabla \oplus \nabla'$ defined on $E \oplus F$ satisfies

$$(\nabla \oplus \nabla')(e_i \oplus f_j) = e_k \otimes \gamma^k{}_{\mu i}\, dX^\mu \oplus f_l \otimes \Gamma'^l{}_{\mu j}\, dx^\mu. \tag{7.11}$$

The resulting curvature is just the direct sum of the curvature of E and F, as implied by equation (7.11).

For tensor product bundles, the first thing to do is to look at the natural connection ∇'' which takes its value on $E \oplus F$. This connection is actually given by the formula

$$\begin{aligned}\nabla''(s \otimes s') &= (\nabla \otimes 1 + 1 \otimes \nabla')(s \otimes s') \\ &= \nabla(s) \otimes s' + s \otimes \nabla'(s').\end{aligned}$$

The curvature of ∇'' reads

$$\nabla'' = g \otimes 1 + 1 \otimes g' \tag{7.12}$$

What about torsion on pullback bundles? Some definitions are in order. We begin with a map $f : M \to F$ and a connection ∇' on the vector bundle E' over F. The resulting natural pullback connection is thus $\nabla = f^* \nabla'$ with Christoffel symbols which are none other than the of ∇'. More precisely

$$\Gamma^i{}_{\mu j} = \Gamma'^i{}_{\alpha j} \frac{\partial x'^\alpha}{\partial x^\mu}.$$

The curvature of ∇ is the pullback of the curvature of ∇':

$$g^i{}_{j\mu\nu} = \frac{1}{2} g^{ij}{}_{j\alpha\beta} \left(\frac{\partial x'^\beta}{\partial x^\mu} \frac{\partial x'^\alpha}{\partial x^\nu} - \frac{\partial x'^\alpha}{\partial x^\nu} \frac{\partial x'^\beta}{\partial x^\mu} \right).$$

Because of their central role in quantization of Chern-Simons-Witten theories, we pause to say a few words about the curvature of projected connections. First, we write the projection as $p : F \to E$ with E a sub-bundle of M. As before, one thinks of ∇ as a connection on F. The projected connection ∇^p on E is

$$\nabla^p(s)\, p\, (\nabla(s)),$$

where as we have seen before, s is a section of F belonging to the sub-bundle E. A peculiar fact is that the curvature ∇^p may be non-trivial, even though the curvature of ∇ may be zero. The Levi-Civita connection on $S^2 \hookrightarrow \mathbb{R}^3$ provides an illustration of this.

7.5   Connections On Principal Bundles

A principal P-bundle is a fiber bundle whose fiber and transition functions both belong to the same matrix group. The gauge potentials of Maxwell's theory of Electromagnetism and Yang-Mills gauge theories are identifiable with connections on principal bundles. The purpose of this present section is to provide a detailed presentation of connections on principal bundles. We shall begin with parallel-transport. Choose a local trivialization with coordinates (x, g) for the principal bundle P, where $g \in G$, G is a matrix group, $\mathcal{G}$ its Lie algebra and $g^{-1} dg$ the Maurer-Cartan form, i.e. a matrix of one-forms belonging to $\mathcal{G}$. For completeness we note that the Maurer-Cartan equations are

$$d\theta = \frac{1}{2}[\theta \wedge \theta] \, d\theta_i = \frac{1}{2} \lambda_i{}^{kj} \, \theta_k \wedge \theta_j; \, d\theta' = \frac{1}{2}[\theta' \wedge \theta'].$$

By a local section of P, we mean a smooth map from a neighborhood U to G. As we have explained earlier, assigning a connection on P provides a rule for the parallel-transport of sections. In view of this, a connection A on P is a Lie algebra valued matrix of 1-forms in $T^*(M)$:

$$A(x) = A^a{}_\mu(x) \frac{\lambda_a}{2i} \, dx^\mu. \tag{7.13}$$

Let $x(t)$ denote a curve in M; the section $g_{ij}(t)$ can be defined to be parallel-transported along x if the following differential equation is satisfied:

$$g_{ik} + A_{\mu ij}(x) \cdot x^\mu \, g_{jk} = 0, \tag{7.14}$$

with A_μ the connection on the principal bundle P. Equation (7. 14) can be rewritten in a more explicit form, namely:

$$g^{-1} \frac{dg}{dt} + g^{-1} \left(A^a{}_\mu(x) \frac{\lambda_a}{2i} \frac{dx^\mu}{dt} \right) g = 0.$$

This concludes the discussion for the parallel-transport case. The focus is now on the tangent space approach.

Recall that parallel-transport along a curve parallel-transport along a curve $x(t)$ allows one to compare the fibers of a principal bundle P at various points of $x(t)$. Drawing on the techniques used for vector bundle connections, we lift curves $x(t)$ in M to curves in P. This procedure requires us to differentiate along a lifted curve. This is actually done by

$$
\begin{aligned}
d/dt &= \dot{x}^\mu \frac{\partial}{\partial x^\mu} + \dot{g}_{ij} \frac{\partial}{\partial g_{ij}} \\
&= \dot{x}^\mu \left(\frac{\partial}{\partial x^\mu} - A^a{}_\mu(x) \frac{(\lambda^a)_{ik}}{2i} g_{kj} \frac{\partial}{\partial g_{ij}} \right) \\
&= \dot{x}^\mu \left(\frac{\partial}{\partial x^\mu} - A^a{}_\mu(x) \right) \bar{L}_a;
\end{aligned}
$$

here, we have implicitly used the parallel-transport equation for g_{ij}. This yields the covariant derivative

$$
D_\mu = \frac{\partial}{\partial x^\mu} - A^a{}_\mu(x) \bar{L}_a. \tag{7.15}
$$

The curvature is

$$
[D_\mu, D_\nu] = -F^a{}_{\mu\nu} \bar{L}_a \tag{7.16}
$$

where

$$
F^a{}_{\mu\nu} = \partial_\mu A^a{}_\nu - \partial_\nu A^a{}_\mu + f_{abc} A^b{}_\mu A^c{}_\nu. \tag{7.17}
$$

As for the cotangent bundle approach, we define the connection on P to be a $\mathcal{G}$-valued one-form ω in $T^*(P)$, whose vertical component is the Maurer-Cartan form $g^{-1} dg$. Locally, we write

$$
\omega = g^{-1} A g + g^{-1} dg,
$$

with

$$
A(x) = A^a{}_\mu(x) \left(\frac{\lambda_a}{2i} \right) dx^\mu.
$$

Notice that A remains invariant under the right action of the group $g \to g_0$ while ω transforms tensorially, i.e.

$$
\omega \to g_0^{-1} \omega g_0.
$$

The resulting curvature is a Lie algebra valued matrix 2-form:

$$
\Omega = d\omega + \omega \wedge \omega = g^{-1} F g; \tag{7.18}
$$

with

$$
F = dA + A \wedge A = \frac{1}{2} F_{\mu\nu}{}^a \frac{\lambda_a}{2i} dx^\mu \wedge dx^\nu.
$$

Equation (7.18) satisfies the Bianchi identity

$$d\Omega + \omega \wedge \Omega - \Omega \wedge \omega = 0. \tag{7.19}$$

According to previous sections, the transition functions of a principal bundle act on fibers by left multiplication. Suppose we have two overlapping neighborhoods U and U' and a transition function $\Phi_{UU'} = \Phi$. It then follows that the local fiber coordinates g and g' in U and U' are related by $g' = \Phi g$. In order for us to write down the connection 1-form in the overlapping region $U \cap U'$, we first require A to transform as

$$A' = \Phi A \Phi^{-1} + \Phi \, d\Phi^{-1}. \tag{7.20}$$

From there, we show without much difficulty that

$$\omega = g^{-1} A g + g^{-1} dg = g'^{-1} A' g' + g^{-1} dg',$$

which in essence is the statement that ω is well defined in $T^*(P)$. The transformation (7.19) is referred to as a gauge transformation of A. By contrast, the gauge transformation for M', a submanifold of M is

$$M'' = \Phi M' \Phi^{-1}.$$

We point out that the curvature 2-form Ω is also well-defined over the manifold i.e.

$$\Omega = g^{-1} M' g = g'^{-1} M'' g'.$$

What are the gauge transformations for pullback bundles? To answer this, we begin by choosing a section $g = g(x)$ by which one can pull back ω and Ω to the base space. The procedure in question tells us that A and M' are equivalent to the pullbacks $g^*\omega$ and $g^*\Omega$. The gauge transformations of A and M' are then simply the changes of the sections.

Drawing on analogies with gauge field theories, we referred to the gauge group as the structure group G; the choice of $G = U(1)$ [1], for instance, gives the theory of Electromagnetism; $G = SU(3)$ [2], gives the theory of strong interactions or Quantum Chromodynamics; $G = E_8 \otimes E_8$ [3], the theory of heterotic superstring. In these particular instances, the (pullback) curvature M' gives the strength of the gauge field. Associated vector bundles act to describe matter field contents in the gauge theory.

By taking a particular connection on the $U(1)$ principal bundle over the base space S^2 and by restricting it to satisfy Maxwell's equations, the resulting physical system corresponds to Dirac's magnetic monopole. A $U(1)$ connection 1-form reads

$$\omega = \begin{cases} A_+ - d\psi_+ & \text{on } D_+ \\ A_- + d\psi_- & \text{on } D_-, \end{cases}$$

where $D_\pm$ are the two hemispheres of S^2. The transition function

$$e^{i\psi_-} = e^{in\phi} e^{i\phi_+},$$

implies the gauge transformation

$$A_+ = A_- + n\phi.$$

Gauge potentials which satisfy Maxwell's equations in $\mathbb{R}^3 - \{0\}$ are of the form

$$A_\pm = \frac{n}{2}(\pm\cos\theta)\, d\phi = \frac{n}{2r}\frac{x\, dy - y\, dx}{z \pm r}.$$

The curvature is therefore

$$\begin{aligned} F &= dA_\pm = \tfrac{n}{2}\sin\theta\, d\theta \wedge d\phi \\ &= \tfrac{n}{2r^3}(x\, dy \wedge dz + y\, dz \wedge dx + z\, dx \wedge dy). \end{aligned}$$

As for the monopole charge, it corresponds to a negative value of the first Chern number c_1 characterizing the bundle:

$$-c_1 = -\int_{S^2} c_1 = \frac{1}{2\pi} F = \frac{1}{2\pi}[\int_{D_+} F_+ + \int_{D_-} F_-] = n.$$

Let us repeat the procedure for the SU(2) Yang-Mills instanton. The fiber is $G = \text{SU}(2) = S^3$ with base S^4. The metric reads

$$ds^2 = \frac{dx_\mu\, dx_\mu}{(1 + r^2/a^2)^2} = \frac{dr^2 + r^2\left(\sigma_x^2 + \sigma_y^2 + \sigma_z^2\right)}{(1 + r^2/a^2)^2} = \sum_{a=0}^{3}(e^a)^2;$$

it is obtained by projection from the north or south pole onto $\mathbb{R}^4$. Using the fact that the overlapping region $D_+ \cap D_-$ is equivalent to S^3 we relate the SU(2) fibers by the transition function

$$g_- = [h(x)]^k \cdot g_+,$$

where k is an integer, $h = \frac{(t - i\lambda \cdot x)}{r}$, and λ are the $SU(2)$ Pauli matrices. The connection 1-form is

$$\omega = \begin{cases} g_+^{-1} A g_+ + g_+^{-1} dg_+ & \text{on } D_+ \\ g_-^{-1} A' g_- + g_-^{-1} dg_- & \text{on } D_-. \end{cases}$$

Notice that A' can be written as

$$A'(x) = h^k(x) A(x) h^{-k}(xy) + h^k(x) dh^{-k}(x).$$

The case $k = 1$ gives the single instanton solution

$$\begin{aligned} D_+ : A &= \frac{r^2}{r^2 + a^2} \cdot h^{-1} dh \\ &= \frac{r^2}{r^2 + a^2} i\lambda_k \sigma_k. \end{aligned}$$

The gauge transformed solution is hence

$$\begin{aligned} D_- : A' & h \left[\frac{r^2}{r^2 + a^2} h^{-1} dh \right] h^{-1} + h\, dh^{-1} \\ &= -\frac{dh\, h^{-1}}{1 + r^2/a^2} \\ &= \frac{i\lambda_k \bar{\sigma}_k}{1 + r^2/a}. \end{aligned}$$

The field strength in $D_\pm$ follows from

$$\begin{aligned} D_- : F_+ &= dA + A \wedge A \\ &= i\lambda_k \frac{2}{a^2} \left(e^0 \wedge e^k + \tfrac{1}{2} \epsilon_{kij}\, e^i \wedge e^j \right); \end{aligned}$$

$$\begin{aligned} D_+ : F_- &= dA' + A' \wedge A' \\ &= h F_+ h^{-1}. \end{aligned}$$

Indeed F is self-dual, that is, $\star F = F$; consequently, the Bianchi identity implies that the Yang-Mills equation

$$D_A \star F = d \star F + A \wedge \star F - \star F \wedge A = 0,$$

is indeed well defined. The instanton number k can be written as

$$\begin{aligned} k &= -c_2 = -\int_{S^4} c_2 = -\frac{1}{8\pi^2} \int_{S^4} Tr\, F \wedge F \\ &= -\frac{1}{8\pi^2} \left[\int_{D_+} Tr\, F_+ \wedge F_+ + \int_{D_-} Tr\, F_- \wedge F_- \right] \\ &= -\frac{1}{8\pi^2} \left(-\frac{48}{a^4} \right) \int_S^4 e^0 \wedge e^1 \wedge e^2 \wedge e^3 \\ &= +1. \end{aligned}$$

7.6 References

[1] Cowan, E.: **Basic Electromagnetism**, Academic Press, New York 1968.

[2] Yndurain, F. Y.: **Quantum Chromodynamics: An Introduction to the Theory of Quarks and Gluons**, Springer-Verlag, 1983 Berlin and New York.

[3] Green, M. B., Schwarz, J. H. and Witten, E.: **Superstring Theory**, Cambridge University Press, 1987 Cambridge and New York.

Chapter 8

Geometric Quantization of Chern-Simons-Witten Theories

A classical physical system can be described by the Poisson algebra of functions on the phase space $\mathcal{A}$. The standard Poisson bracket,

$$\{f,g\} = \frac{\partial f}{\partial p}\frac{\partial g}{\partial q} - \frac{\partial f}{\partial q}\frac{\partial g}{\partial p} \tag{8.1}$$

is associated with Hamiltonian mechanics, in particular with dynamical system of the type:

$$\begin{aligned} \dot{q} &= \tfrac{\partial H}{\partial p}(p,q) \\ \dot{p} &= -\tfrac{\partial H}{\partial q}(p,q), \end{aligned}$$

with q_i the generalized coordinates and p_i the momenta defined in the phase space $\mathcal{A} = \mathbb{R}^{2n}$. The resulting Poisson algebra or Lie-Poisson bracket reads:

$$\{f,g\} = \left\langle x_{ij}, \left[\frac{\partial f}{\partial x_i} \times \frac{\partial g}{\partial x_j}\right] \right\rangle . \tag{8.2}$$

Quantization is the procedure by which one associates to each theory a Hilbert space $\mathcal{H}$ of quantum states, and a map m from a subset of the Poisson algebra to the space of symmetric operators on $\mathcal{H}$.

Geometric quantization is essentially a globalization of canonical quantization in which the additional structure needed for quantization is explicitly expressed in geometric terms. According to a classical theorem by Darboux [1], the space of differential k-forms on the space of smooth real functions satisfies the Jacobi identity:

$$\{\{f,g\},k\} + \{\{g,k\},f\} + \{\{k,f\},g\} = 0 \qquad (8.3)$$

This usually allows one to define a Poisson manifold, together with a Poisson algebra $\mathcal{F}$. In light of this, the Poisson bracket in (8.1) is in reality a bivector field Ψ that belongs to the space of bivector fields on $\mathcal{A}$. Put differently,

$$\{f,g\} = \Psi(df, dg).$$

Note that Ψ stands for the mapping of bundles

$$\Psi : T^*(\mathcal{A}) \to T(\mathcal{A}),$$

which is linear on the fibers. We can therefore give a reformulation of the initial Poisson bracket:

$$\{f,g\} = \langle df, \Psi \, dg\rangle \, ;$$

the main novelty here is the angles, which denote the pairing of a form and a vector field. In terms of local coordinates $x_1, \cdots, x_{2n}$ on $\mathcal{A}$, we have

$$\{f,g\} = \Psi_{jk}\, \partial_j f \, \partial_k g, \qquad (8.4)$$

where $\partial_j = \frac{\partial}{\partial x_j}$. The Jacobi identity, written in terms of the tensor components of Ψ_{jk}, is of the type

$$\sum_{(j,l,m)} \Psi_{jk}\, \partial_k \Psi_{lm} = 0. \qquad (8.5)$$

Note that the summation is done only over cyclic permutations. The rank $r(\Psi) = \operatorname{rank} \Psi(x)$ is in general smaller than the dimension of $\mathcal{A}$. If $r(\Psi)$ is a constant, then we say of the Poisson bracket on $\mathcal{A}$ that it is of constant rank. To say that the bracket (8.4) is nondegenerate mean that $r(\Psi) = \dim \mathcal{A}$. This corresponds to the case for which the dimension of the manifold $\mathcal{A}$ is necessarily even. $\mathcal{A}$ is then referred to as a symplectic manifold [1, 2, 3, 4], and its corresponding closed, nondegenerate 2-form ω

$$\omega = \frac{1}{2}\,\Psi\,(x)^{-1}\sum_{i=1}^{n}\,dx_i\,\wedge\,dx_j, \qquad (8.6)$$

is called a symplectic form.

The use of symplectic manifolds as phase spaces in physics is well-documented [1, 2, 3], and additionaly [12]. In classical mechanics, the trajectories,

$$\ddot{x}_i = -\frac{\partial v}{\partial x_i},$$

are determined by the values of x and $\dot{x}$ at $t = 0$. The Lagrangian

$$\mathcal{L} = \int\left(\frac{1}{2}\,\dot{x}^2 - v(x)\right)\,dt$$

or

$$\mathcal{L} = \int\left(\dot{p}x - (\frac{1}{2}\,p^2 + v)\right)\,dt$$

encompasses most solutions to the problem of finding the adequate trajectories. The momentum p, an independent variable in these Lagrangians, is equivalent to $\dot{x}$ for classical orbits. The classical phase space can be thought of as the space of classical solutions of the equations of motions. A classical solution would thus be determined by the initial values x and $\dot{x}$. The symplectic manifold $\mathcal{A} = \mathbb{R}^{2n}$ is the phase space with symplectic form

$$\omega = dp \wedge dx, \qquad (8.7)$$

which, in view of our above discussion, is equivalent to (8.6).

Although ω is closed, it is not necessarily exact. For compact manifolds of dimensions $2n > 0$ for instance, ω would not be exact or the volume form $\omega \wedge \cdots \wedge \omega$ cannot exact and the volume would be zero (which of course is absurd). As a consequence, compact symplectic manifolds must admit non-contractible 2-cycles. A celebrated instance of the lack of this particular property is that of S^4: that is why S^4 has no symplectic structure whatsoever [1, 2, 3, 4].

8.1 Holomorphic and Symplectic Structures on Bundles

This section is entirely devoted to the study of various aspects of gauge theory on Riemann surfaces which are relevant to the geometric quantization of Chern-Simons-Witten or topological quantum field theories. We begin with a $\mathbb{C}^\infty$ vector bundle $\mathcal{L}$ over Σ_g, a Riemann surface of genus g. A connection is a differential operator

$$d_A : \Omega^0 \left(\Sigma_g; \mathcal{L} \right) \to \Omega^1 \left(\Sigma_g; \mathcal{L} \right), \tag{8.8}$$

such that

$$d_A \left(fs \right) = df \otimes s + f \, d_A s,$$

where f is a $\mathbb{C}^\infty$ function, $s \in \Omega^0 \left(\Sigma_g; \mathcal{L} \right)$ a section, and $\Omega^p \left(\Sigma_g; \mathcal{L} \right)$ denotes p-forms on Σ_g with values in $\mathcal{L}$. The local form of the connection is

$$d_A = d + A = d + b \, d\bar{z} + c \, dz,$$

where A is a matrix of 1-forms and b, c are matrix-valued functions.

The fundamental invariant of a connection is its curvature, ∇_A. If we extend d_A to p-forms, then

$$\nabla_A = d_A^2 : \Omega^0 \left(\Sigma_g; \mathcal{L} \right) \to \Omega^2 \left(\Sigma_g; \mathcal{L} \right). \tag{8.9}$$

In local terms, (8.9) reads

$$\begin{aligned}
\nabla_A \left(s \right) &= \left(d + A \right) \left(d + A \right)(s) \\
&= d^2 s + dAs - Ads + Ads + A^2 s \\
&= \left(dA + A^2 \right) s.
\end{aligned} \tag{8.10}$$

In other words, the curvature is linear over $\mathbb{C}^\infty$-functions and can be considered as an element of $\Omega^2 \left(\Sigma_g, \operatorname{End} \mathcal{L} \right)$ or, a matrix valued 2-form with respect to (8.10).

A gauge transformation is an automorphism of the bundle $\mathcal{L}$. At the local level, a gauge transformation g is a $\mathbb{C}^\infty$-function with values in GL $(n, \mathbb{C})$. Gauge transformations act on connections by conjugating the differential operator

$$g^{-1} \, d_A g.$$

Again, there is a local interpretation of this formula, namely:

$$g^{-1} d_A g = g^{-1} (d + A)g = dg^{-1} (dg) + g^{-1} Ag. \tag{8.11}$$

For the curvature to vanish, a necessary and sufficient condition is for the connection to be flat. The vanishing of the curvature is the integrability condition for the local existence of n-linearly independent solutions of the equation

$$d_A (s) = 0.$$

Let $(s_1, \cdots, s_n)$ and $(\bar{s}_1, \cdots, \bar{s}_n)$ be such two bases of solutions with

$$\bar{s}_i = \sum_j a_{ij} d_A s_j.$$

The relation follows that

$$\begin{aligned} d_A \bar{s}_i &= d_A \left(\sum a_{ij} s_j \right) \\ &= \sum da_{ij} \otimes s_j + \sum a_{ij} d_A s_i \\ &= \sum da_{ij} \otimes s_j \\ &= 0, \end{aligned}$$

where the a_{ij} are constant matrices. Thus, a flat connection gives rise to a family of constant transition functions for $\mathcal{L}$, which in turn defines a representation of the fundamental group $\pi_1 (\Sigma_g)$ into GL $(n, \mathbb{C})$ [5].

8.1.1 Holomorphic Structures on $\mathcal{L}$

A holomorphic structure on $\mathcal{L}$ is a differential operator

$$d_b'' : \Omega^0 (\Sigma_g; \mathcal{L}) \rightarrow \Omega^{0,1} (\Sigma_g; \mathcal{L}) \tag{8.12}$$

satisfying the relation

$$d_b'' (f, s) = d'' f \otimes s + f d_b'' s,$$

with $d'' f = \frac{\partial f}{\partial \bar{z}} d\bar{z}$. In local terms

$$d_b'' = d'' + b \, d\bar{z},$$

where b is a matrix-valued function.

Two holomorphic structures are said to be equivalent if there is a gauge transformation g such that

$$g^{-1} d''_{b_1} g = d''_{b_2},$$

is a holomorphic isomorphism. A connection on $\mathcal{L}$ defines a holomorphic structure providing that we put

$$d''_A s = (d_A s)^{0,1}.$$

Consider now a local trivialization of the bundle $\mathcal{L}$ endowed with Hermitian structure. We have the form

$$d'' = d'' + b d\bar{z},$$

and a connection

$$d_A = d + b d\bar{z} - b^* dz.$$

This connection is compatible with the Hermitian structure since two sections s, t of $\mathcal{L}$ satisfy

$$d < s, t > = \langle d_A s, t \rangle + \langle s, d_A t \rangle. \tag{8.13}$$

We refer to (8.13) as a unitary connection. Conjugating by a unitary gauge transformation (i.e. one which preserves the Hermitian inner product), takes a unitary connection to a unitary connection. Hence, we draw the conclusion that the space of holomorphic structures on $\mathcal{L}$ and the space of unitary connections are similar objects.

Let $\Sigma_g \times \mathbb{C}$ denote the trivial line bundle. $\mathbb{C}^\infty$ complex line bundles are determined topologically by their first Chern class or degree. To classify holomorphic line bundles, a standard approach is thus to look at bundles of degree zero. On the trivial line bundle $\Sigma_g \times \mathbb{C}$, every holomorphic structure is equivalent to the holomorphic structure of a flat unitary connection. These connections, it should be noted, are unique modulo gauge transformations. This implies that one can parametrize holomorphic line bundles of degree zero by unitary equivalence classes of flat connections. As noted earlier, these equivalence classes are in fact classes of constant transition functions.

In the abelian case, they correspond to the cohomology group

$$H^1 (\Sigma_g; U(1)) \simeq \frac{H^1 (\Sigma_g; \mathbb{R})}{H^1 (\Sigma_g; \mathbb{Z})},$$

a $2g$-dimensional real torus, the Jacobian. This object will play a central role in the geometric quantization of topological quantum field theories below. To endow this torus with a complex structure, the use of Quillen's determinant bundle [6] is required.

8.1.2   Stability of Holomorphic Bundles

A rank two holomorphic bundle $\mathcal{L}$ of degree zero is said to be stable whenever every sub-bundle $\mathcal{L}' \subset \mathcal{L}$ has degree less than zero. This is a natural condition which is unfortunately obscured by its cumbersome derivation using geometric invariant theory. As a general rule, a holomorphic structure on $\mathcal{L}$ originating from a flat, unitary and irreducible connection implies that $\mathcal{L}$ is stable. Thus, flat connections give rise to stable holomorphic structures. The theorem which gives to the flat connections their central role is known as the Narasimhan-Seshadri theorem [8]. Roughly, it states that every holomorphic structure on a given vector bundle over Σ_g of degree zero which is stable is indeed equivalent to the holomorphic structure of a flat, irreducible unitary connection. The connection is unique modulo unitary gauge transformations.

One immediate consequence of the Narasimhan-Seshadri theorem is that the space of equivalence classes of (stable rank n) bundles of degree zero is in one to one correspondence with the equivalence classes of irreducible representations $\pi_1(\Sigma_g) \to U(n)$ modulo the action of conjugation by $U(n)$. The resulting space is a Hausdorff space owing to the fact that $U(n)$ is compact. This space plays a crucial role in some forms of geometric quantization of Chern-Simons-Witten theories, as noted by Witten et al. [7] and Atiyah [9].

8.1.3   Symplectic Geometry and Gauge Transformations

A symplectic manifold $\mathbb{R}^{2n} = \mathcal{A}$ is a manifold endowed with a nondegenerate closed 2-form ω. Two basic examples of symplectic manifolds are:

1. The cotangent bundle $T^\star(M)$ of a manifold M [1, 2, 3];

2. A Kähler manifold M [2, 4].

Symplectic geometry originated from Hamiltonian mechanics. Nowadays it is a powerful tool in shedding light on the relation between flat connections and stability. A symplectic manifold $\mathcal{A}$ admits the action of a Lie group G, which preserves the symplectic form. This can be illustrated as follows. Let v denote a vector field generated by such an action. It leaves ω fixed, meaning that the Lie derivative of ω is zero:

$$\mathcal{L}_v(\omega) = d\left(i\left(v\right)\omega\right) + i\left(v\right)d\omega = 0.$$

Since ω is closed, we thus have the relation

$$d\left(i\left(v\right)\omega\right) = 0,$$

implying that for $H^1\left(\mathcal{A}; \mathbb{R}\right) = 0$, there exists a function μ_v such that

$$d\mu_v = i\left(v\right)\omega.$$

There is such a function for each vector field generator of the Lie group G. More precisely, for each element of the Lie algebra $\mathcal{G}$, the (dual) linear map follows

$$\mathcal{G} \to \mathbb{C}^\infty\left(\mathcal{A}\right),$$

or equivalently

$$\mu : \mathcal{A} \to \mathcal{G}^\star. \tag{8.14}$$

The map μ is referred to as a moment map [7, 9] whenever μ commutes with the natural action of G on $\mathcal{A}$ and $\mathcal{G}^\star$. When G preserves the symplectic form and the metric (i.e. the complex structure in brief) in $\mathcal{A}$, it then follows that $\mathcal{G} \otimes \mathcal{G}$ yields a complex Lie algebra of holomorphic vector fields. This comes mostly from the action of a complexication $G_{\mathbb{C}}$ of G on the symplectic phase space $\mathcal{A}$ [1, 2, 3, 4, 7, 9]. This situation is related to the stability of points in $\mathcal{A}$, providing that they suitably transform under $G_{\mathbb{C}}$ to points on the zero set of the moment map (8.13). Below, we analyze this stability for two finite-dimensional examples. Stable points are essential in the framework of geometric quantization if one is to avoid the occurence of global anomalies. I shall defer to Chapter 9 an explanation of how unstable or degenerate points in symplectic manifolds give rise to global anomalies.

Back to the two examples. In the first one, we take the manifold M to be the space of $n \times n$ complex matrices with Kähler metric $\mathrm{Tr}\left(A\,A^\star\right)$.

It is required here that $G = U(n)$ act by conjugation. The moment map corresponding to this case is

$$\mu\left(A\right) = \frac{1}{2} i\left[A, A^\star\right], \qquad\qquad (8.15)$$

where we have identified $\mathcal{G}$ and $\mathcal{G}^\star$ by the invariant form $\mathrm{Tr}\left(AA^\star\right)$. The stable points are those which are conjugate by $\mathrm{GL}\left(n, \mathbb{C}\right)$ to a matrix in the zero set of μ, namely, a normal matrix. Every normal matrix can, in principle, be diagonalized by a unitary transformation, and consequently the stable matrices are the diagonalizable ones.

In the second example, we consider $M = \mathbb{C}P^1 \times S^2$ and set $G = \mathrm{SO}(3)$. The moment map is simply the inclusion of the unit sphere in $\mathbb{R}^3$. For $M = \mathbb{C}P^1 \times \mathbb{C}P^1 \times \mathbb{C}P^1 \times \mathbb{C}P^1$, the set of ordered quadruple points is $\mathbb{C}P^1$. The zero set of the moment map consists of those quadruples whose center of mass is at the origin. Thus, the stable points are those which transform under $\mathrm{PSL}\left(2, \mathbb{C}\right) = \mathrm{SO}(3)^{\mathbb{C}}$.

Our focus is now on the infinite dimensional case. The Kähler manifold M is now taken to be the space of holomorphic structures on the bundle $\mathcal{L}$. This space is essentially an affine space, the difference of two structures being an element $b \in \Omega^{0,1}\left(M; \mathrm{End}\,\mathcal{L}\right)$. Tangent vectors are given by $\dot{b} \in \Omega^{0,1}\left(M; \mathrm{End}\,\mathcal{L}\right)$. We also specify the Kähler metric

$$i \int_M \mathrm{tr}\left(\dot{b}^\star \dot{b}\right).$$

As for the group G, it now corresponds to the group of unitary gauge transformations $G_{\mathbb{C}}$, the group of all complex gauge transformations. In order for us to extract the stable points, we have to find the moment map. As an initial step, let us think of M as being the vector space of unitary connections. Its corresponding symplectic form is

$$\omega\left(\dot{a}, \dot{b}\right) = \int_M \mathrm{tr}\left(\dot{a} \wedge \dot{b}\right), \qquad\qquad (8.16)$$

with $\dot{a}, \dot{b} \in \Omega^{0,1}\left(M; \mathrm{End}\,\mathcal{L}\right)$. The Lie algebra of G consists of the skew-Hermitian section, $\Psi \in \Omega^0\left(M; \mathrm{End}\,\mathcal{L}\right)$ for which the action

$$e^{-t\Psi} d_a e^{t\Psi},$$

yields a vector field corresponding to Ψ:

$$\dot{a} = d_a \Psi \ (= s\Psi + [a, \Psi] \text{ locally}).$$

Thus, using Stokes's theorem, we find

$$d\mu_\Psi (\dot{a}) = \int_M \text{tr} \, (d_a \Psi \wedge \dot{a}) = \int_M \text{tr} \, (\Psi \, d_a \dot{a}). \qquad (8.17)$$

Hence, $d_a \dot{a} = d\dot{a} + [a, \dot{a}]$ locally. However, the curvature $\nabla_a = da + a^2$, so

$$\dot{\nabla}_a = d\dot{a} + \dot{a}a + a\dot{a} = d_a \dot{a},$$

and consequently, the moment map μ is given by

$$\mu(a) = \nabla_a \in \Omega^2 (M; \text{End} \, \mathcal{L}). \qquad (8.18)$$

The stability of points allows us to see rather beautifully that the stable holomorphic structures are those which are equivalent to the zero-set of μ, i.e. the flat unitary connections. Any consistent geometric quantization of topological quantum field theories has to take this information into account. A powerful observation from the present discusssion is that the Narasimhan-Seshadri theorem established a correspondence between the infinite-dimensional stability of holomorphic bundles with that of the finite dimensional one. This observation plays a central role in the construction of the Donaldson invariants for 4-manifolds. But it is beyond the scope of this book, let alone this chapter, to discuss further this particular subject.

8.1.4 Generalized Gauge Transformations on Symplectic and Holomorphic Structures

The action of the group of gauge transformations on the space of connections leads us, via the moment map, to zero curvature equations, and places them in a natural and general context. We shall use this method to describe more generalized set of equations, and to see, along the way, their relevance in the description of holomorphic structures on M. We start by considering the cotangent bundle $T^* M$ of the space of holomorphic structures on the trivial bundle $\mathcal{L} = M \times \mathbb{C}^2$. The tangent vectors to M are given by $\dot{a} \in \Omega^{0,1} (M; \text{End } \mathcal{L})$, and so the cotangent vectors are elements of $\Phi \in \Omega^{0,1} (M; \text{End } \mathcal{L})$ under the pairing:

$$\int_M \text{tr} \left(\Phi \wedge \dot{a} \right).$$

The manifold $T^* M$ is symplectic as an infinite-dimensional complex manifold: the full group $\mathcal{G}_{\mathbb{C}}$ of complex automorphisms of $\mathcal{L}$ acts on it, preserving the symplectic form, and this in turn, yields a moment map.

Next, we consider the action of the Lie algebra $\Omega^0 (M; \text{End } \mathcal{L})$ of $\mathcal{G}_{\mathbb{C}}$ on $T^* M$. Since the action of an automorphism g can be written as

$$\left(g^{-1} d_b'' g, g^{-1} \Phi g \right),$$

a given element $\Psi \in \Omega^0 (M; \text{End } \mathcal{L})$ will generate a vector field on $T^* M$, i.e.

$$\left(\dot{b}, \dot{\Phi} \right) = \left(d_b'' \Psi, [\Phi, \Psi] \right). \tag{8.19}$$

The following natural symplectic form is given by (8.19)

$$\omega \left((\dot{b}_1, \dot{\Phi}_1), (\dot{b}_2, \dot{\Phi}_2) \right) = \int_M \text{tr} \left(\dot{\Phi}_1 \, b_2 - \Phi_2 \, b_1 \right). \tag{8.20}$$

Consequently,

$$\begin{aligned} d\mu_\Psi \left(\dot{b}, \dot{\Phi} \right) &= \int_M \text{tr} \left([\Phi, \Psi] \, \dot{b} - \dot{\Phi} \, d_b'' \Psi \right) \\ &= \int_M \text{tr} \left(\Psi \left(d_b'' \dot{\Phi} + \left[\dot{b}, \Phi \right] \right) \right), \end{aligned}$$

and this gives the moment map

$$\mu \left(b, \Phi \right) = d_b'' \Phi. \tag{8.21}$$

The zero set of this complex moment map is the set of holomorphic structures d_b'' and sections Φ of $\text{End } \mathcal{L} \otimes K$, which are holomorphic with respect to d_b''. This zero set can be endowed with a complex submanifold of $T^* M$. Let its induced Kähler metric be

$$\| (\dot{b}, \dot{\Phi}) \|^2 = i \int_M \text{tr} \left(\dot{b}^* \dot{b} + \dot{\Phi} \Phi^* \right)$$

and consider the moment map for the group $\mathcal{G}$ of unitary automorphisms.

The action on holomorphic structures (or similarly, unitary connections) is the curvature ∇_A. The conjugation action of unitary automorphisms on $\Phi \in \Omega^{1,0}(M; \text{End}\,\mathcal{L})$ corresponds to the finite-dimensional example of $U(n)$ acting on the complex matrices as above. The result is the moment map

$$\nabla_a + [\Phi, \Phi^\star]. \tag{8.22}$$

In combination with the complex moment map, this gives the natural set of equations

$$\begin{aligned} d''_a \Phi &= 0 \\ \nabla_a + [\Phi, \Phi^\star] &= 0, \end{aligned} \tag{8.23}$$

for a unitary connection a and section $\Phi \in \Omega^{1,0}(M; \text{End}\,\mathcal{L})$. Note that equations (8.23) are the self-dual Yang-Mills equations for connection over $\mathbb{R}^4$.

8.2 Geometric Quantization of Chern-Simons-Witten Theories

The original discovery by Edward Witten of 3-manifold invariants in 1988 [10] required, in its Hamiltonian version, the geometric quantization of the space of flat connections on a compact surface Σ_g. According to Atiyah and Bott [5], the manifold M which corresponds to the space of gauge equivalence classes of flat G-connections can be regarded as a symplectic quotient of the space of all connections. The canonical symplectic form admits various values of the level k (which, incidentally, correspond to different symplectic structures.) These symplectic manifolds are canonically associated to Σ_g. To quantize them requires the choice of a Kähler polarization, and one then needs to prove -for consistency purposes- that the resulting space is independent of that choice. Let us pause a bit to explain what we mean by Kähler polarization. We pick a symplectic manifold, (M, ω) and consider its integral class $\frac{1}{2\pi}[\omega] \in H^2(M; \mathbb{R})$. Next, we choose a line bundle $\mathcal{L}$ with unitary connection whose curvature is ω. To choose a complex structure on M such that ω is Kähler means that one must choose the $(0,1)$ part of the covariant derivative ∇ of the connection in question in such a way that it gives a holomorphic structure on $\mathcal{L}$. The corresponding projective space is the quantization relative to the polarization. To complete the quantization, one shows that the projective space is, in a suitable sense, independent of the choice of polarization. It is worth pointing out the following: as M acquires a complex structure, the Narasimhan-Seshadri theorem transforms it into the moduli space of stable holomorphic vector bundles over Σ_g.

We begin with the geometric quantization of

8.2.1   Canonical Spaces

Let $\mathcal{A} = \mathbb{R}^{2n}$ denote an affine symplectic space with a symplectic, nondegenerate 2-form,

$$\omega = \omega_{ij}\, d\, a_i \wedge da_j,$$

and coordinates $a_1, a_2, a_{2n-1}, a_{2n}$. The 2-form ω is viewed as a transformation $T \to T^*$, where T is the tangent space to $\mathcal{A}$ and T^* its dual. We write $\omega_{ij} = \omega^{-1} : T^* \to T$ for the inverse of ω. The constraints are

$$\omega \frac{\partial}{\partial x_k} = \omega_{kj}\, dx_j$$
$$\omega^{-1}\, dx_k = \omega_{kj}\, \frac{\partial}{\partial x_j}.$$

The symplectic form ω can be reformulated as

$$\omega = \sum_{i=1}^{n} a_i \wedge a_j. \tag{8.24}$$

To $\mathcal{A}$, we associate a Hilbert space $\mathcal{H}(\mathcal{A})$ or $\mathcal{H}_{\mathcal{A}}$; it is the Hilbert space of all square-integrable functions of $a_1, \cdots, a_{2n}$. The coordinates a_i act on $\mathcal{H}_{\mathcal{A}}$. The Heisenberg commutation relations

$$[a_i, a_j] = -i\omega_{ij}, \tag{8.25}$$

yield the Heisenberg algebra.

The primary objective of quantizing $\mathcal{A}$ is to produce a Hilbert space. In reality though, the quantization procedure for $\mathcal{A}$ gives an irreducible unitary Hilbert space of representation of the algebra of (8.25). Roughly speaking, we will construct a representation of the Heisenberg group. This group, it turns out, is an extension by $U(1)$ of the group of affine translations of $\mathbb{R}^{2n}$. According to the Stone-von Neumann uniqueness theorem, there is a unique such

representation. The theory furthermore implies that the projective space of
$\mathcal{H}_{\mathcal{A}}$ is independent of the choice of the symplectic coordinates a_i, a_j; as a
consequence of this fact, we may regard $\mathcal{H}_{\mathcal{A}}$ as a projective representation
of the symplectic group. We will construct the Heisenberg algebra in great
detail. For now, however, let us pause to say that there exists another way
to construct $\mathcal{H}_{\mathcal{A}}$ using mostly complex coordinates. As a general rule, this
space can be identified with the space of polynomials in

$$\alpha_i = a_i + i a_j.$$

In actual fact though, a complex structure on $\mathcal{A}$ has to be chosen in such a
way that the symplectic form originates from a Hermitian metric. So we begin
by picking a unitary line bundle $\mathcal{L}$ over $\mathcal{A}$ with a connection ∇ such that the 2-
form curvature is $\tau^2 = -i\omega$. According to Axelrod, Della Pietra and Witten
whose work appears in [7], such a bundle exists whenever $\frac{\omega}{2\pi}$ represents an
integral cohomology class. Since the isomorphism class $H^1\left(\mathcal{A}; U(1)\right)$ of $\mathcal{L}$ is
trivial, it then follows that $\mathcal{L}$ is unique up to isomorphism. In the process,
this verifies the uniqueness which, we may recall, comes as a requirement
from the Stone-von Neumann uniqueness theorem.

To obtain an irreducible representation of the Lie algebra, we must choose
a complex structure c on $\mathcal{A}$; then a second step requires that the symplectic
2-form ω be positive and compatible with c, or, put differently, that in c,
ω be a $(1,1)$ form. Combining these requirements yields holomorphic linear
functions α_i which are defined by the formula

$$\omega = i \sum_i d\alpha_i^\star \wedge d\bar\alpha_i.$$

Once c is picked, $(\mathcal{L}, \nabla)$ acquires a new structure: ∇ has curvature $(1,1)$
with respect to c, so ∇ and c combine to give $\mathcal{L}$ a holomorphic structure.
Explicitly, the holomorphicity of $\mathcal{L}$ originates from the condition

$$\bar\partial^2 \left(\nabla\nabla\right)^{0,2} = \omega^{0,2} = 0,$$

where $\bar\partial$ is a $\nabla^{0,1}$ operator on $\mathcal{L}$.

We are in a position to give a detailed description of $\mathcal{L}$. To this end, let
$\mathcal{L}_0$, endowed with the Hermitian metric, denote the trivial holomorphic line

bundle on $\mathcal{A}$. Furthermore, let $e \in \mathbb{C}^\infty (\mathcal{L}_0)$. The relations follows

$$|e|^2 = e^* e \exp - b$$
$$b(\alpha) = \sum \alpha_i \alpha_i^{-1}.$$

From results in reference [7], we deduce that the curvature of $\mathcal{L}_0$ is

$$-\bar{\partial}\partial b = \sum d\alpha_i \, d\bar{\alpha}_i$$
$$= -i\omega. \tag{8.26}$$

There is an isomorphism between $\mathcal{L}$ and $\mathcal{L}_0$ which is unique up to projectivity and is a direct consequence of (8.26).

What is the Heisenberg algebra to which we alluded earlier? It has a simple expression,

$$[\alpha_i, \bar{\alpha}_j] = -\delta_{ij}$$
$$[\alpha_i, \alpha_j] = [\bar{\alpha}_i, \bar{\alpha}_j] \tag{8.27}$$
$$= 0,$$

where, we may recall, δ_{ij} is the Kronecker delta (see Chapter 5). The form (8.27) has the following presentation in the Hilbert space $\mathcal{H}_\mathcal{A}$:

$$\beta(\alpha_i) \cdot e = \alpha_i \, e$$
$$\beta(\bar{\alpha}_i) \cdot e = \frac{\partial}{\partial \alpha_i}. \tag{8.28}$$

For the commutation relations to be well-defined, the following constraints must be taken into consideration:

$$\frac{\partial}{\partial \alpha_i}(\alpha_j e) = \delta_{ij} e + \alpha_j \frac{\partial}{\partial z_i} e.$$

This representation of the Heisenberg group is irreducible for the simple reason that most holomorphic functions can be expressed in terms of polynomials.

We have the Siegel upper half space (which was studied in Chapter 3 and 4), the space of admissible complex structures on $\mathcal{A}$. As such, one can vary c while keeping ω positive -and of type $(1,1)$- while preserving simultaneously the translational invariance of c. We write S for the Siegel upper half space. As we pointed out in Chapters 3 and 4, S is the homogeneous space $Sp(2n; \mathbb{R})/U(n)$ of complex symmetric $n \times n$ matrices with positive imaginary part. Let $\mathcal{H}_S$ denote a bundle of the Hilbert space over S; the symplectic group $\mathrm{Sp}(2n; \mathbb{R})$ acts on $\mathcal{H}_S$. The fiber over $c \in S$ is just $\mathcal{H}_c$. Now, using the Stone-von Neumann uniqueness theorem for the representation of Heisenberg groups, one finds that the identification of all of the fibers of $\mathcal{H}_S$ comes from endowing $\mathcal{H}_S$ with a natural projectively flat connection. Below we discuss in great detail the nature of this form.

The first priority then concerns the expansion of the commutators. They are

$$[\nabla_{ij}, \nabla_{kl}] = -i \left(\nabla_{il}\, \omega_{jk} + \nabla_{ik}\, \omega_{jl} + \nabla_{lj}\, \omega_{ik} + \nabla_{kj}\, \omega_{il} \right).$$

This gives the Lie algebra of $\mathrm{Sp}(2n; \mathbb{R})$, ξ, or more precisely, the algebra of homogeneous quadractic polynomials under Poisson bracket. One would expect ξ to act in any representation of the Heisenberg algebra. Notice that, as a group, $\mathrm{Sp}(2n; \mathbb{R})$ acts on the Heisenberg algebra by outer automorphism. In doing so, it conjugates the representation $\mathcal{H}_c$ of the Heisenberg group to another representation. $\mathrm{Sp}(2n; \mathbb{R})$ acts projectively on $\mathcal{H}_c$ by the uniqueness of (this) irreducible representation. Consequently, ∇_{ij} gives the action at the level of ξ.

Note that the ∇_{ij} are second-order differential operators (the a_i act on $\mathcal{H}_c$ as zeroth-order differential operators and the $\bar{\alpha}_i$ act on as first order differential operators.) Vector fields generated by the ∇_{ij} act transitively on the Siegel upper space S. They define, in turn, a connection on the Hilbert space $\mathcal{H}$ and this connection is given by a second-order differential operator which is inherited from the ∇' s:

$$\nabla^{1,0} = \delta^{1,0} + \frac{1}{4} \nabla \left(\delta c \circ \omega^{-1} \nabla^{1,0} \right). \tag{8.29}$$

To extend the quantization over the whole Hilbert space requires the use of the Teichmüller space $\mathcal{T}$. Essentially, this is done by using the period mapping [7, 9] as defined by equations (8.13-14-17).

8.2.2   The Torus

This case corresponds to the geometric quantization of Chern-Simons-Witten theories for the gauge group $U(1)$. Let us begin by considering an integer lattice

$$\Lambda = \mathcal{L}_{\mathbf{Z}} \subset \mathcal{A}_{\mathbb{R}},$$

of maximal dimension $2N$, corresponding to the quotient torus

$$T = \mathcal{A}/\Lambda \simeq \mathbb{R}^n/\mathbb{Z}^n = \mathcal{A}(U(1)). \qquad (8.30)$$

As was the case previously, our interest lies in the complex quantization case. To this effect, we pick, once again, a complex structure c on $\mathcal{A}$. The torus, as defined in formula (8.30), can then be regarded as an abelian variety, $\mathcal{A}_c$. The complex line bundle on $\mathcal{A}$, namely $\mathcal{L}$ with curvature $2\pi i\omega$, descends to a holomorphic line bundle on $\mathcal{A}_c$, with first Chern class ω.

An important issue centers around the fact that the line bundle $\mathcal{L}$ is not well-defined, mainly because the torus is not simply connected [5, 7]. There exist however, several ways to get rid of this ambiguity in $\mathcal{L}$. We refer the interested reader to Atiyah's book [9, pages 20-22], to reference [7], and [12].

The action of Λ commutes with the connection ∇. Hence, ∇ can be shown to restrict to a connection on the sub-bundle $\mathcal{H}_\Lambda$. This implies that ∇ can be made flat by tensoring $\mathcal{H}_\Lambda$ with a suitable line bundle whose connection lie in S. A short note of caution though: the resulting connection will not necessarily be invariant under the action of the group $\mathrm{Sp}(2n;\mathbb{Z})$; this is so because the original connection differs by a central factor. This is the most probable cause of anomalies generated by an analogous problem: the need to rescale the connection 1-form in the infinite dimensional case from the normalization it would otherwise have in finite dimension. Very little is known about these class of anomalies, except some indications that the Atiyah-Patodi-Singer Index theorem may be of potential use in their detection and cancellation, as noted in [7].

8.2.3   The Symplectic Quotient

The lattice is replaced by a compact Lie group G which acts symplectically on $\mathcal{A}$. There is a lift of the G-action to an action on the line bundle $\mathcal{L}$, preserving the connection ∇. In what follows, we shall refocus the original quantization of the affine spaces toward the G-invariant complex structures. Consider the Siegel space S_G to which we associate the Hilbert space bundle $\mathcal{H}_G$ whose fiber over c is the G-invariant subspace $(\mathcal{H}_c)_G$ of $\mathcal{H}_c$. To obtain a natural projectively flat connection on $\mathcal{H}_G$, we only need restrict our attention to S_G and $(\mathcal{H}_c)_G$.

As a symplectic manifold, the quotient $\mathcal{A}_c/\mathcal{G}_c$ is independent of the complex structure c. We write $\mathcal{A}//\mathcal{G}$ to denotes the symplectic or Marsden-Weinstein quotient of $\mathcal{A}$. It should be noted that $\mathcal{A}//\mathcal{G}$ acquires a complex structure from its identification with $\mathcal{A}_c/\mathcal{G}_c$. The connection (8.29) is once again given by a second-order differential equation.

8.2.4 Non-Abelian Moduli Space of Representations

The method under consideration here can be regarded as a generalization of the previous torus case. Consider a compact, simply connected Lie group with corresponding Lie algebra κ. Associated to it is an oriented compact surface without boundary, Σ_g. Let A' be the space of connections on a principal G-bundle, $P \rightarrow \Sigma_g$; this is an affine space whose gauge group $\mathcal{G}$ acts on A' by the classical transformation

$$d_A \rightarrow g d_A g^{-1}.$$

Write (Σ_g, G) for the resulting space, i.e. a truly Hausdorff space of dimension $2(n-1) \cdot \dim G$. The Lie algebra valued 1-forms α and β define the skew pairing by

$$\{\alpha, \beta\} \cong \frac{1}{4\pi} \int_{\Sigma_g} (\alpha \wedge \beta), \tag{8.31}$$

or more generally for $G = \mathrm{SU}(n)$:

$$\{\alpha, \beta\} = \int_{\Sigma_g} -\mathrm{Tr}(\alpha \wedge \beta). \tag{8.32}$$

Equations (8.31-32) combine to define a natural symplectic structure on A'. The group of gauge transformations $\mathcal{G}$ acts naturally on A' preserving its symplectic structure.

The next problem is to find the curvature. This means reformulating the moment map (8.14-15-18) as

$$m : A' \rightarrow \mathcal{G}.$$

Consequently, the curvature reads

$$m(A') = \Omega_{A'}.$$

$\Omega_{A'}$ is a Lie algebra valued 2-form. With these ingredients in hand, let us pick a symplectic form ω_0 on A', an integral class obtained as the curvature of a line bundle over A'. The resulting symplectic quotient is

$$\mathcal{M} = A'//\mathcal{G} = \{A' : \Omega_{A'} = 0\}/\mathcal{G}. \qquad (8.33)$$

The Narasimhan-Seshadri theorem [8], which was interpreted by Atiyah and Bott [5] as an analogue for the infinite-dimensional affine space of connections, tell us that once a complex structure c is picked on Σ_g, the moduli space $\mathcal{M}$ has a natural identification with the moduli space of holomorphic principal $\mathcal{G}_c$-bundles on Σ_g. In actual fact though, the family of complex structures on $\mathcal{M}$ is parametrized by the Teichmüller space, $\mathcal{T}$. Thus, the quantization result in building a second-order differential operator, that is, projectively flat connections on the bundle $\mathcal{L}$ over $\mathcal{T}$.

8.3 Symplectic Quantization of 2-Dimensional Surfaces

The geometric (or symplectic) quantization of a two-dimensional surface is challenging. Among the obstacles encountered, the lack of polarization (which gives the direction of projection) in the phase space $\mathcal{A}$, and the absence of a primitive θ for the symplectic form ω are the most noticeable ones. Thus, it is of no surprise that the ordinary rules of geometric quantization fail here. This is best exhibited by the obstruction in defining the cohomology class $[\mu] \in H^1(\mathcal{A}; \mathbb{Z})$. As shown in previous sections, the need for a (Kähler) polarization and a well-defined cohomology class $[\mu] \in H^1(\mathcal{A}; \mathbb{Z})$ are ingredients in any consistent quantization of Chern-Simons-Witten theories.

In order for us to put these obstacles in a manageable form, we consider the framework in which Stokes's theorem evolves. This theorem suggest, among other things, that integrals of the 2-form ω over $\Sigma_g \subset \mathcal{A}$ with boundary $\partial \Sigma_g = M$ can be regarded as analogues of the integrals of a 1-form θ over one-dimensional cycles. To properly extend the analogy to a one-dimensional integer-valued class $[\mu]$, we need the equivalent of an integer-valued 2-form on the entire symplectic phase space $\mathcal{A}$. Just in the same way as the differential form ω represent the double-cohomology class $[\omega]$, this sought integer-valued

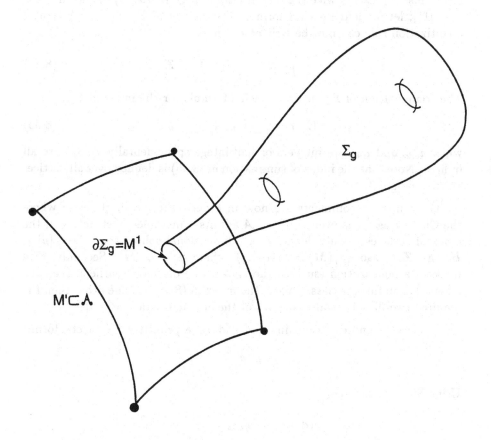

Figure 8.1:

form must represent twice the Chern class $2c_1 \in H^2(\mathcal{A}; \mathbb{Z})$. Following Turaev [11], let us define such a form as the index i of Σ_g, $i(\Sigma_g)$. A typical quantization rule can now be written down as

$$\frac{1}{2\pi i} \int_{\Sigma_g} \omega - \frac{1}{4} i(\Sigma_g) \in \mathbb{Z}. \qquad (8.34)$$

The triangulation of Σ_g yields a specific formula for the index of Σ_g:

$$i(\Sigma_g) = \sum i_{\delta\beta\alpha} + \sum \mu_{\beta\alpha}, \qquad (8.35)$$

where $i_{\delta\beta\alpha}$ and $\mu_{\beta\alpha}$ are integers or half-integers; incidentally, α, β, δ are all indices. Note that the index of summation in (8.35) is defined over all vertices of the triangulation.

The following comments are now in order. Firstly, in the case where the Chern class c_1 is even (that is, $\mathcal{A}$ admits a metaplectic structure), the index of surfaces usually defines a one-dimensional cohomology class, $[\mu] \in H^1(M'; \mathbb{Z}_4)$; also, $[\mu](M) = i_M(\Sigma_g)$, where $\partial\Sigma_g = M'$. Secondly, if $\mathcal{A}$ posseses a polarization (such as the Kähler one discussed earlier), then the class $[\mu]$ is an integer class. Hence, the index in (8.35) is none other then the required two-dimensional analogues of the characteristic class $[\mu]$.

Let $\{\mu_{\beta\alpha}\}$ denotes a cochain on $\mathcal{A}$ and $c_{\beta\alpha}$ a primitive symplectic form:

$$c_{\beta\alpha} = \int_{c_i} \theta_\alpha - \int_{c_i} \theta_\beta.$$

Using Stokes's theorem, we find

$$(dc)_{\delta\beta\alpha} + \Phi_{\delta\beta\alpha} = \int_{\Sigma_g} \omega.$$

In light of this, the quantization rule (8.34) needs to be reconsidered; its new form is

$$\frac{1}{h} \Phi_{\delta\beta\alpha} - \frac{\pi}{2} i_{\delta\beta\alpha} = a_{\alpha\delta} + a_{\delta\alpha} + a_{\beta\alpha} - 2\pi \, p_{\delta\beta\alpha}; \qquad (8.36)$$

here $p_{\delta\beta\alpha}$ denotes an integer-valued 2-cocycle while $\{a_{\beta\alpha}\}$ stands for a certain 1-cochain on $\mathcal{A}$.

The following relation holds:

$$d\left(a_{|M'} + \frac{1}{h} a - \frac{\pi}{2}\mu\right) = 2\pi p_{|M'}.$$

Thus, the 1-cocycles with values in $U(1)$

$$\lambda_{\beta\alpha} = \exp\left(i\left(a_{\beta\alpha} + \frac{1}{h}c_{\beta\alpha} - \frac{\pi}{2}\mu_{\beta\alpha}\right)\right), \tag{8.37}$$

together with the cohomology class $[\lambda] \in H^1\left(M'; U(1)\right)$ lives on M'. There is, however, a subtlety: the geometry of $[\lambda]$ is not all that transparent from this approach. Furthermore, the contribution of the symplectic form as $h \to 0$ is mixed with the contribution of the index in equation (8.37). It is with these issues in mind that we proceed to take a closer look at $[\lambda]$.

On the boundary $\partial\Sigma_g \subset M$, the exact form for $[\lambda]$ is

$$[\lambda]_{\partial\Sigma_g} = \exp\left(\frac{i}{h}\int_{\Sigma_g}\omega - i\frac{\pi}{2}i_M\left(\Sigma_g\right)\right). \tag{8.38}$$

This result arises by actually writing the integral of the form ω in terms of the cocycle Φ and the cochain c, that is,

$$\int_{\Sigma_g}\omega = \sum \Phi_{\delta\beta\alpha} + \sum c_{\beta\alpha}.$$

Thus,

$$\begin{aligned}
\frac{1}{h}\int_{\Sigma_g}\omega - \frac{1}{2}i_M\left(\Sigma_g\right) &= \sum\left(\frac{1}{h}\Phi_{\delta\beta\alpha} - \frac{\pi}{2}i_{\delta\beta\alpha}\right) \\
&+ \sum\left(\frac{1}{h}c_{\beta\alpha} - \frac{\pi}{2}\mu_{\beta\alpha}\right) \\
&= -2\pi\sum a_{\delta\beta\alpha} \\
&+ \sum\left(p_{\beta\alpha} + \frac{1}{h}c_{\beta\alpha} - \frac{\pi}{2}\mu_{\beta\alpha}\right).
\end{aligned} \tag{8.39}$$

Combining (8.39) with (8.37) give

$$\begin{aligned}
\exp\left(\frac{i}{h}\int_{\Sigma_g}\omega - i\frac{\pi}{2}i_M\left(\Sigma_g\right)\right) &= \exp\left(i\sum\left(a_{\beta\alpha} + \frac{1}{h}c_{\beta\alpha} - \frac{\pi}{2}\mu_{\beta\alpha}\right)\right) \\
&= [\lambda]_{\partial\Sigma_g} \\
&= \prod \lambda_{\beta\alpha}.
\end{aligned} \tag{8.40}$$

We are now in position to state the following important result which removes the mentioned obstacles to quantizing two-dimensional surfaces: *Any two-dimensional surface that admits a quantization rule of the type given by equation (8.34) must have a trivial (or vanishing) $[\lambda] \in H^1\left(M; U(1)\right)$.*

8.4 References

[1] Arnold, V. I. and Givental', A. B. : **Symplectic Geometry** Springer-Verlag, (1990) Berlin New York.

[2] Fomenko, A. T. : **Symplectic Geometry**, Gordon and Breach, (1988) New York.

[3] Weinstein, A. : *Symplectic Manifolds and their Lagrangians Submanifolds*, Adv. in Math. 6 (1971) 329-346.

[4] Weinstein, A. : **Lectures on Symplectic Manifolds**, Conf. Board Math. Sciences Regional Conf. Ser. Math. Vol. 29 (1979) Amer. Math. Society, Providence, R. I.

[5] Atiyah, M. F. and Bott, R. : *The Yang-Mills Equations Over Riemann Surfaces*, Trans. Roy. Soc. Lond. A 308 (1982) 523-612.

[6] Quillen, D. : *Determinants of Cauchy-Riemann Operators Over a Riemann Surface*, Funct. Anal. and Applications 19 (1985) 31-34.

[7] Axelrod, S., Della Pietra, S. D. and Witten, E. : *Geometric Quantization of Chern-Simons Gauge Theory*, J. Diff. Geometry 33 (1991) 787-902.

[8] Narasimhan, M. S. and Seshadri, C. S. : *Stable and Unitary Vector Bundles on a Compact Riemann Surface*, Ann. Math. 82 (1965) 540-567.

[9] Atiyah, M. F. : *The Geometry and Physics of Knots*, Cambridge Univ. Press, 1990 Cambdridge, New York.

[10] Witten, E. : *Quantum Field Theory and the Jones Polynomial*, Comm. Math. Phys. 121 (1989) 351-399.

[11] Turaev, V. G. : *Cocycles for Symplectic First Chern Class and Maslov Indices*, Funct. Anal. Applications 18 (1984).

[12] Murayama, H.: *Explicit Quantization of the Chern-Simons Action*, Z. Phys. C-Particles and Fields 48 (1990) 79-88.

Chapter 9

Deformation Quantization

In the present chapter, we will analyze the relation between two seemingly disparate subjects: global anomalies and deformation quantization. The framework for this draws on the work by Baadhio, in reference [1], which asserts that under suitable conditions, global anomalies arise when deformation quantization is performed. We shall learn that in the theory of deformation quantization, there exist essentially two appraoches which may induce the occurrence of global anomalies: the first one has to do with an obstruction to patching a locally deformation quantizable $\star$-product to a global $\star$-product. Most Poisson manifolds are known to admit regular, or otherwise said, local deformation quantizations. Their generalization, however, represents a substantial problem involving topological obstructions and the like. We will not explicitly discuss the case of global anomalies induced by this technique, the reason being that they have not being investigated at the time of writing. We will, however, address the problem of global anomalies which arise under the degeneracy of the symplectic 2-form ω. In particular, we will study the case in which ω degeneracy extends to the degeneracy of the Poisson structure on the classical phase space, $\mathcal{A} = \mathbb{R}^{2n}$. What one finds is that singularities arise which strongly restrict the possibility of carrying a globalization of local deformations quantization. This is much like the patching approach to globalizing local deformation quantization.

We begin with a background review of deformation quantization.

9.1   Aspects of Deformation Quantization

From chapter 8, we have learned that quantization is a procedure consisting of associating a Hilbert space $\mathcal{H}$ of quantum states to a given classical theory. In view of Dirac's correspondence principle [2], quantization of a classical system is the process of associating Hermitian operators on a Hilbert space to the classical observables. These are functions

$$f \in \mathbb{C}^{\infty}(\mathcal{A});$$

where $\mathcal{A} = \mathbb{R}^{2n}$, a symplectic manifold, denotes the classical phase space. The operation, up to a constant factor depending on the Planck constant $\hbar$, associates to the Poisson bracket on $\mathcal{A}$,

$$\{f, g\} = \left\langle \xi, \left[\frac{\partial f}{\partial \xi} \times \frac{\partial g}{\partial \xi} \right] \right\rangle,$$

the commutator of the operators. The resulting operator $\hat{f}$ associated with f is called the quantization of f.

The problem of quantization is then to construct an associative non-commutative algebra of quantum observables satisfying the correpondence principle outlined in [2], and futhermore, to describe its representations by operators in the Hilbert space of quantum states. The issue, as one might guess, is a highly non-trivial one under Dirac's correspondence principles. The reason for this is in fact simple and inescapable: we are asking that the quantization produce similar observables for classical and quantum quantum systems alike. But observables in quantum mechanics, unlike those in classical mechanics, do not commute with one another! The study of deformation (and geometric) quantizations was originally intended to deal with this dichotomy. One mathematical transcription of Dirac's principle is that the obtained representation of the Poisson algebra, $\mathbb{C}^{\infty}((\mathcal{A})\{,\})$ be irreducible. A theorem of van Hove [3] shows, however, that the quantization has no such solution. In order to avoid this contradiction, physicists often quantized in two steps:

• the pre-quantization which consists of linear representations of the whole algebra $\mathbb{C}^{\infty}((\mathcal{A})\{,\})$ by operators on a given complex vector space. Then, in a second phase,

• the actual quantization, where one restricts the problem to a convenient sub-algebra of C^∞ $((\mathcal{A})\{,\})$, and then one represents its *irreducibility* (in a way that is compatible with the underlying physical theory) on a Hilbert space initially built out of the pre-quantization space.

Kostant [4] found, a while ago, that modulo a certain obstruction, the pre-quantization problem could be solved by means of the space of cross-sections of a complex line bundle,

$$\pi : \mathcal{L} \rightarrow \mathcal{A},$$

the so-called pre-quantization bundle, and, by a well-defined pre-quantization formula. The existence of $\mathcal{L}$ is actually related to whether the cohomology class of the symplectic 2-form ω is an integral cohomology class, i.e.

$$[\omega] \in H^2(\mathcal{A}, \omega).$$

9.1.1 Star-Product as Deformation Quantization

The theory of deformation quantization was developed a decade ago by Lichnerowicz [5] et al. It originated in the Dirac principle that a quantum system has to reduce to the corresponding classical system as $\hbar \rightarrow 0$ ($\hbar = h/2\pi$, h = Planck constant). We take $\hbar$ to be the numerical value of the Planck constant when it is expressed in a unit of action characteristic of the class of systems under consideration. This present formulation avoids the paradox that we consider the limit $\hbar \rightarrow 0$ even though Planck's constant is a fixed physical parameter. The quantization of a classical system, in order to satisfy the correspondence principle, should therefore consist of a deformation of this classical system into a system that depends on a parameter, $\hbar$.

It is important that we define what we mean by system. According to our presentation of geometric quantization in the previous chapter, from the classical standpoint, the word system refers to the commutative algebra of observables, $C^\infty(\mathcal{A}) \otimes \mathbb{C}$, e.g. the algebra of smooth functions on the classical space $\mathcal{A}$ with ordinary multiplication, and the Poisson bracket $\{,\}$ on $\mathcal{A}$, induced by the symplectic 2-form ω.

The quantization of this system consists of defining an associative, noncommutative deformation of the usual product, into a new operation, the

$\star$-product ,

$$f \star g \ (f, g) \in \mathbb{C}^{\infty} \ (\mathcal{A})$$

which depends on $\hbar$, and is such that the commutant,

$$[f \star g] \overset{\text{def}}{=} \frac{(f \star g - g \star f)}{i\hbar},$$

will be a deformation of the Lie algebra operation $\{f, g\}$. From a purely mathematical standpoint, the goal of deformation quantization is to construct $\star$-product and interpret the resulting physics. The undeformed product $\star_0$ is taken to represent the usual pointwise multiplication, so that $(\mathcal{A}, \star_0)$ is the algebra of classical observables. In applying Dirac's general principles [2], the limit

$$\lim_{\hbar \to 0} \left[(a \star_\hbar b - b \star_\hbar a) / i\hbar \right],$$

is equivalent to a given classical Poisson bracket $\{a, b\}$ on the phase space $\mathcal{A}$. This bracket is a Poisson structure in the sense that it can be shown to satisfy the axioms of a Lie algebra, together with the Leibniz identity, namely, $\{ab, c\} = \{a, c\} b + a \{b, c\}$. In this context, a *formal deformation* $D = D_0, D_1, \cdots$ is called a $\star$-product whenever each of the bilinear map D_i is indeed a differential operator, annihilating the constant functions when $i \geq 1$. These conditions are essential in ensuring the localness of the $\star$-product, while simultaneously ensuring that the constant function 1 remains as the unit element.

To recapitulate what we have said so far. On the classical phase space $\mathbb{C}^{\infty} \ (\mathcal{A} = \mathbb{R}^{2n})$ of smooth functions, we have two basic algebraic structures:

- the associative product,

$$\mathcal{A} \times \mathcal{A} \to \mathcal{A} : (a, b) \to ab;$$

- a Poisson bracket associated to the symplectic nondegenerate 2-form ω,

$$\mathcal{A} \times \mathcal{A} \to \mathcal{A} : (a, b) \to \{a, b\}.$$

A $\star$-product is a deformation of the associative structure, which, by antisymmetrization, gives rise to a deformation of the Poisson bracket $\{a, b\}$. We call this procedure a deformation quantization. It originated with the need to verify Dirac's correpondence principles [2] between classical and quantum mechanics. Deformation quantization and $\star$-product are essentially similar objects and from now on, we will use these two terms interchangeably. The symplectic form ω, expressed in coordinates $(q_1, \cdots, q_n, p_1, \cdots, p_n)$, reads

$$\omega = \sum_i dq_i \wedge dp_i,$$

and the Poisson structure,

$$\{a, b\} = \sum_j \left(\frac{\partial a}{\partial q_j} \frac{\partial b}{\partial p_j} - \frac{\partial a}{\partial p_j} \frac{\partial b}{\partial q_j} \right),$$

is invariant under all diffeomorphisms preserving ω, so there is a well-defined Poisson structure on any symplectic manifold.

The most readily and extensively studied example of a $\star$-product is the so-called Moyal-Weyl product [6] on $\mathbb{R}^{2n}$, with the Poisson structure just described above. The Moyal-Weyl product originates from the composition of operators on $\mathbb{C}^\infty(\mathbb{R}^n)$ via Weyl's identification [7] of such operators with functions on $\mathbb{R}^{2n}$ and was used by Moyal in reference [6] to investigate quantum statistical mechanics. Let V denote a vector space, and consider α, a skew-symmetric bilinear function on $V^\star$. The formula

$$\{a, b\} = \alpha(da, db)$$

defines a Poisson structure on v. Associated to the bilinear operator, α, is a unique differential operator:

$$\Xi : \mathbb{C}^\infty(V \times V) \to \mathbb{C}^\infty(V \times V)$$

with constant coefficients for which

$$\{a, b\} = \Delta^\star \Xi(a \otimes b).$$

The term $a \otimes b$ is the function $(y, z) \mapsto a(y) b(z)$. Moreover, we think of $\Delta^\star$ as

$$\Delta^\star : \mathbb{C}^\infty(V \times V) \to \mathbb{C}^\infty(V).$$

We can define the Moyal-Weyl product on V by

$$a \star_\hbar b = \Delta^\star \exp\left(\frac{i\hbar\Xi}{2} \right) (a \otimes b).$$

The space $\mathbb{C}^\infty(V)[\hbar]$ associated with this product is called the Weyl algebra of v.

9.1.2 Local and Global Characters of Deformation Quantization

On a given Poisson manifold, the Leibniz identity implies that the Poisson bracket is given by a skew-symmetric contravariant tensor (or bivector) field α, which is called Poisson tensor. It is given by the formula

$$\{a, b\} = \alpha\,(da, db).$$

A case of interest is the one in which the rank of α (that is, the rank of the matrix function $\alpha_{kl}\,(x)\,\{x_k, x_l\}$ in local coordinates) is constant. A theorem by Lie [8] is then applied, and roughly speaking, states that the Poisson manifold $\mathcal{A}$ is locally isomorphic to a vector space with constant Poisson structure. Thus, Poisson manifolds which are labelled as regular are always locally deformation quantizable [9]. Now, there is a highly non-trivial problem: patching together the local deformations to produce a global $\star$-product. The issue of giving a global character to the local deformation quantization centered around some forms of obstructions which are especially responsible for the occurrence of quantum pathologies, the so-called global anomalies. The interested reader may want to consult [13] for a thorough investigation of obstructions and global anomalies, particularly insofar as they relate to the context of ten-dimensional physics.

Under special circumstances, the patching of local quantizations can be done rather easily. Here is how. It begins with the realization that the Moyal-Weyl product on a vector space V, with constant Poisson structure, is invariant under all the affine automorphisms of V, mostly because an operator with constant coefficients is invariant under such transformations. This, in turn, suggests the possibility of constructing a global quantization of any symplectic manifold $\mathcal{A}$, covered by local isomorphisms, and for which the transition maps are affine. Such a covering, it was revealed in [9], exists whenever the phase space $\mathcal{A}$ admits a flat torsionless linear connection (for which the covariant derivative $\nabla \alpha$ is exactly zero).

Torsionless Poisson connections play an important role in the analysis of deformation quantization. In particular, the existence of a deformation quantization in the presence of a flat torsionless Poisson connection was first established in Bayen et al.'s landmark papers on deformation quantization [5]. Note that the consistency condition on the connection (e.g. $\nabla \alpha = 0$) implies the regularity of the Poisson structure. Situations in which this is not the case are those with degenerate Poisson structure (which we shall investigate at length in upcoming sections). At this level, as we shall soon see, quantum pathologies arise and need to be cancelled if the theory is to be unique and consistent.

Let us mention another case of interest concerning the possible globalization of local deformation quantizations. This case is actually the most challenging of all, and research activities aimed at solving it are still very much ongoing. When the Poisson manifold $\mathcal{A}$ does not admit a flat, torsionless Poisson connection, the authors of [5], and independently Gutt in [10], have pointed out the existence of an obstruction localized in the Hochschild cohomology space $H^3(\mathcal{A})$. More precisely, the obstruction to the existence of the local $\star$-product, $D_1, D_2, \cdots$ lies in $H^3(\mathcal{A})$.

However, it was later found that the obstructions could further be pinpointed within the de Rham cohomology class in $H^3(\mathcal{A})$. As a consequence, there is seemingly no obstruction to constructing a deformation quantization when the third Betti number of the phase space $\mathcal{A}$ is zero. One can arrive at this conclusion by simply looking at the much smaller Chevalley cohomology space, $H^3_{\text{chev}}(\mathcal{A})$ (taken as a Lie algebra via the Poisson bracket). Deformations of this Lie algebra were initially studied by Vey in reference [11]. But the proof is credited to De Wilde and Lecomte [12] that deformation quantization always exists on any symplectic manifold. The most celebrated consequence of the De Wilde-Lecomte proof is that it renders trivial, in the symplectic case, the class of obstructions (to any generalization of locally deformed quantization) for $H^3_{\text{deRham}}(\mathcal{A})$ trivial. For manifolds other than symplectic ones, the question remains and, as we mentioned earlier, research is still ongoing.

9.1.3   Issues With Deformation Quantization

Overall, the purpose of this is chapter is to report on the relationship between deformation quantization and global anomalies. Global anomalies occur whenever *large* gauge transformations of a classical field theory fail to be symmetries of the corresponding quantum theory. In this case, the anomalies are referred to as global gauge anomalies. Global gravitational anomalies, on the other hand, arise whenever the effective action of a given theory is non-invariant under a diffeomorphism group that cannot be smoothly deformed to the identity, that is, to mapping class groups; the later condition is essential to prevent the occurrence of disconnected general coordinate transformations, a sure manifestation of anomalies.

Both global gauge and global gravitational anomalies were discovered a decade ago or so by Witten [14]. Since then, the subject has seen a fruitful extension. The case pertinent to the study of global gravitational anomalies in topological quantum field theories [15, 16] has elicited the use of three-dimensional mapping class groups and constitutes such an example. There is also the use of some classical invariants of links and knots (such as the Arf invariant) which have been shown to detect global gravitational anomalies [17]. Finally, although our knowledge of two-dimensional mapping class groups is fairly secure, the recent impetus for studying three-dimensional mapping class groups [18], for which very little was known, originates once again in the newly emerged physical theories in dimension three (including quantum gravity).

In carrying out a standard geometric quantization, we are accustomed to the fact that the symplectic manifold, the phase space $\mathcal{A} = \mathbb{R}^{2n}$, ought to be endowed with a nondegenerate 2-form ω (see chapter 8, and additionally, references [1, 3, 4, 20, 32, 33]). We have come to rely on ω's nondegeneracy property if we are to achieve a flawless quantization. There are several reasons why physicists choose to work with this particular property. The most obvious one owes much to the fact that a degenerate 2-form ω inevitably gives rise to a deformed Poisson bracket; and since deformations of degenerate 2-forms are global functions of a Poisson bracket on $\mathcal{A}$, such a choice renders geometric quantization much less attractive than it would otherwise be.

A Poisson bracket which is defined by a degenerate symplectic 2-form ω will be referred to as a deformed or degenerate Poisson bracket. As we shall learn, the degeneracy of ω essentially implies the occurrence of singularities in $\mathcal{A}$. Under suitable topological conditions, the study of the degeneracy of ω can be reduced to that of singularities in the symplectic leaves Ω_i. A symplectic 2-form ω is said to be degenerate when it is non-zero (or non-trivial), i.e. when it is not exact in the first cohomology class of the symplectic leaf in which it lives.

Nondegenerate Poisson brackets do admit small local perturbations whose harmful effects can be put into a manageable form, as they can be eliminated by perturbation theory. We will provide an example of global anomalies arising from deformed Poisson brackets. The most relevant case is an extension of the (singular) topological structures of the leaves Ω_i. We show in particular that whenever there does not exist a globally smooth homotopy transformation within the Poisson manifold (the collection of all Ω_i defines the Poisson manifold), global anomalies are bound to arise. This situation is reminiscent of global gravitational anomalies arising due to the presence of disconnected general coordinate transformations [14, 15, 16, 17, 18], for we are dealing with transformation of coordinates around singularities in $\mathbb{R}^n$.

The relation between quantization and anomalies, it should be noted, has long been a successful endeavor. Schaller and Schwarz, for instance, have worked out anomalies generated by geometric quantization of fermionic fields [22]. The case of stochastic quantization generating anomalies has been studied by Morita and Kaso in [25], and independently in [23] by Morita. Panfil has investigated the general problem of canonical quantization associated to anomalies [24].

Anomalies induced by the proposed quantization approach of Carlip and Kallosh for the Green-Schwarz superstring have surfaced in [26]. Chiral anomalies coming out of stochastic quantization appeared in a recent work by Kim [27]. There is also the very recently published study of chiral anomaliesof W_3-gravity [28] which are used as a basis to simplify quantization. The emphasis of these works has not been on global anomalies, however.

It will become apparent to the reader that we have not specified which type of global anomalies (i.e. global gauge or gravitational) we are dealing with throughout this chapter. There is a two-fold reason for this. While writing this chapter, we have consistently aimed at establishing the general framework in which global anomalies will manifest themselves once it is determined that we have a deformed quantization. This choice has come at the expense of specifying the type of global pathologies we are dealing with. Secondly, in forthcoming chapters (12 and 13 in particular), we will extensively discuss specific types of global anomalies and the quantum field theories they are known to affect.

9.2   Symplectic Degeneracy of Poisson Brackets

In $\mathbb{R}^3$, the Poisson bracket takes the form

$$\{f,g\} = \left\langle \xi, \left[\frac{\partial f}{\partial \xi} \times \frac{\partial g}{\partial \xi}\right] \right\rangle \qquad (9.1)$$

with $\xi \in \mathbb{R}^3$. The term inside [ ] is a vector product. The 3-form tensor Ψ

$$\Psi(\xi) = \begin{pmatrix} 0 & \xi_3 & -\xi_2 \\ -\xi_3 & 0 & \xi_1 \\ \xi_2 & -\xi_1 & 0 \end{pmatrix}, \qquad (9.2)$$

gives rise to a symplectic form

$$\omega = \omega_\Omega = r\, \sin\theta\, d\theta \wedge d\phi\, (\theta, \psi),$$

where (θ, ϕ) are spherical angles in $\mathbb{R}^3$, and the Ω_i are symplectic leaves. An interesting fact about equation (9.2) is that is it endowed with only one Casimir operator, namely $c(\xi) = |\xi|^2$. This means that the singular point $\xi = 0$ corresponds to a point where the rank of Ψ goes to zero. In other words, $\xi = 0$ is a zero-dimensional singular symplectic leaf, which we write as $\Omega_{|0} = 0$. In $\mathbb{R}^3$, its corresponding algebra is SU(2).

Next, we focus on $\mathbb{R}^4$. We pick coordinates ξ_0, ξ_1, ξ_2, and numbers c_0, c_1, c_2 which satisfy the inequalities $c_0 > c_1 > c_2 > 0$. The resulting Casimir operator for this bracket is similar to that of $\mathbb{R}^3$:

$$c_0(\xi) = \sum_{i=1}^{3} (\xi_i)^2. \qquad (9.3)$$

Equation (9.3) is the perfect example of a degenerate Poisson bracket. It can be generalized by the following formula

$$c_0(\xi) = (\xi_0)^2 + \sum_{i=1}^{3} c_i \cdot (\xi_i)^2. \qquad (9.4)$$

The topological consequences of equations (9.3-4) are of the utmost importance in the investigation of global anomalies. According to these, the symplectic leaves Ω_i are the joint level surfaces of the Casimir operators c_0 and c_1. Thus,

$$\Omega = \{c_0 = \text{constant}; \ c_1 = \text{constant}\}.$$

These leaves are essentially two-dimensional objects since rank of $\Psi = 2$. They also have the following topological structures, to which we shall come back later for a precise study of the manifestations of global anomalies.

$$\Omega \approx S^2 \text{ if } c_1 > c_0\,c_1 > 0 \text{ or } c_0\,c_2 > c_1 > c_0c_3 > 0$$
$$\Omega \approx T^2 \equiv S^1 \times S^1 \text{ if } c_0\,c_1 > c_1 > c_0c_2 > 0.$$

Because the leaves degenerate and become *exactly* zero-dimensional at the points where some of these inequalities turn into an equality, these topological structures are of fundamental interest to us.

9.3 Inducing Global Pathologies

Nondegenerate Poisson brackets admit small local perturbations which are not harmful since they can be eliminated by perturbation theory. In the degenerate case however, we have a very different solution: small or infinitesimal perturbation of the bracket (9.1) yields a global change in the topology of the symplectic manifold $\mathcal{A}$. We shall refer to these deformations as global anomalies. Below, we provide an illustration of global anomalies induced by a degenerate Poisson bracket in $\mathbb{R}^6$.

On $\mathbb{R}^6 = \mathbb{R}^3_M \oplus \mathbb{R}^3_X$ there is a family of linear brackets which depends on a certain parameter $\epsilon \geq 0$:

$$\begin{aligned} \{M_\alpha, M_\beta\} &= M_\gamma \\ \{X_\alpha, X_\beta\} &= \epsilon\,M_\gamma, \end{aligned} \tag{9.5}$$

The α, β and γ are cyclic permutations of the indices $1, 2$ and 3; M and X are subsets of $\mathbb{R}^6$. Altogether, the relations in (9.5) give rise to a bracket on the Lie co-algebra $SO(4)^*$ when $\epsilon > 0$. On the other hand, for $\epsilon = 0$, we have a bracket on the co-algebra $SO(3)^*$. The following Casimir operators

$$c_1 = M \cdot X,$$

and

$$c_2 = |X|^2 + \epsilon\,|M|^2,$$

generate gives rise to terms which are diffeomorphic to $S^2 \times S^2$ for $\epsilon > 0$, and diffeomorphic to $T^*(S^2)$ for $\epsilon = 0$.

As a consequence of this, there does not exit a global, smooth homotopic change of variables in $\mathbb{R}^6$ which can appropriately transform a non-perturbed or deformed SO(3)* bracket into a perturbed SO(4)* one. This observation is the first manifestation of global anomalies.

In quantum field theories, it is often sufficient to know only the first approximation in ϵ if one is to write a pathology-free consistent physical theory. For instance, one may be required to find a Lagrangian transformation of $\mathbb{R}^6$ that maps an SO(3)* bracket into the original and standard Poisson bracket in (9.1). Such a mapping is considered reasonable within $O(\epsilon^2)$ accuracy. A transformation of this type often looks like

$$(m, x) \rightarrow (M^\epsilon, X^\epsilon) = (m, x) + \epsilon v(m, x) + O(\epsilon^2); \qquad (9.6)$$

where

$$v(m, x) = \left(m, \frac{(m \cdot x) m - |m|^2 x}{2|x|^2} \right).$$

Several facts about the point at which the bracket (9.1) degenerates are of interest. Firstly, there are non-singular points in SO(3)* where the rank of (9.1) degenerates. This in turn implies that the generator of the transformation (i.e. the vector field V on $\mathbb{R}^6$) possesses a singularity at $|x| = 0$, i.e. there are nonregular points in SO(3)* where the rank of the bracket drops. So we are lead in a straightforward way to global anomalies.

9.4 Occurrence and Manifestations of Global Anomalies

To begin with, we choose once again our phase space to be the symplectic manifold $\mathcal{A} = \mathbb{R}^{2n}$; its corresponding Poisson bracket is of the form

$$\{f, g\} = \Psi_{jk} \, \partial_j f \, \partial_k g \; = \; < df, \, \Psi \, dg >, \qquad (9.7)$$

where we have used $\partial_j = \frac{\partial}{\partial \xi_j}$; also, the angle brackets denote the pairing of a form and a vector field. We note that Ψ belongs to the space of bi-vector fields; incidentally, Ψ is a mapping of bundles, i.e.

$$\Psi : T^*(\mathcal{A}) \to T(\mathcal{A}).$$

A global anomaly corresponds to forms for which the operation

$$\Psi(df, dg) + \epsilon \Psi(df, dg) \tag{9.8}$$

is a Poisson bracket mod $O(\epsilon^2)$ corresponding to solutions of the equation

$$[\Psi, \Phi] = 0. \tag{9.9}$$

We point out that

$$\sum_{j,l,m} \Psi_{jk} \, \partial_k \Psi_{lm} = 0, \tag{9.10}$$

is the Jacobi identity.

This established, our next order of priority is to distinguish between local and global anomalies. Local anomalies have trivial local deformations which are nonetheless solutions of (9.9). Local anomalies will arise whenever we deal with a family of smooth transformations,

$$\xi \mapsto \varsigma = \xi + \epsilon v(\xi) + O(\epsilon^2), \tag{9.11}$$

in a neighborhood of any point on $\mathcal{A}$. These anomalies typically behave by transforming the initial Poisson bracket (9.1) into the (anomalous) Poisson bracket (9.8). By using a step-by-step approximation with respect to $O(\epsilon^2)$, one can show that the effect of the local anomalies is harmless, for they can give a smooth, homotopic, non-anomalous SO(3) transformation into an anomalous SO(4) one. Their local character is preserved as long as they do not generate non-smooth homotopic transformations. Distinguishing between those two classes of anomalies is what we shall refer to as Problem 1.

In Problem 2, the issue of interest is to find a way to detect the global trivial deformations in $\mathcal{A}$ of equation (9.7).

Problem 3 shall focus on determining the numbers of generators that are invariant under the Poisson bracket (9.1).

Another focus of interest, which will be covered in Problem 4, is to relate the class of non-locally trivial and non-globally trivial deformations to the topology of the symplectic leaf Ω. In actuality, we may assume, without loss of generality, that non-local deformations do assume a global pathological character, whereas non-global deformations are simply local anomalies coming from (9.11). However, to clearly differentiate between local and global anomalies in this framework, we need a better way of clarifying these two situations.

In Problem 5, we will be concerned with global anomalies arising from the Jacobi identity (9.10). As we shall learn, these anomalies are present whenever the condition (9.9) is violated. To describe the anomalies in this case will require a good look at the class of 3-tensors B for which the condition

$$[\Psi, \Phi] = B \tag{9.12}$$

can be solved with respect to Φ. In due course, we will see that the consistency condition (9.12) distinguish between local and global anomalies, thus solving the issues spelled out in Problem 4.

Finally, in Problem 6, we aim to find all solutions to the condition (9.12) in order to effectively cancel global anomalies which arise in deformation quantization. There, we will come upon a generalized version of the descent equations, also known as the Wess-Zumino consistency condition.

9.5 $\star$-Product and Anomalies

Consider a dynamical system

$$\dot{\varsigma} = (\Psi(\varsigma) + \epsilon\,\Phi(\varsigma))\,\partial H(\varsigma), \tag{9.13}$$

where H stands for the Hamiltonian. This system is relevant to Problems 1-5 for it is a system in which the Poisson structure is perturbed instead of the (usual) Hamiltonian of the theory. The most relevant perturbing term in (9.13) is $\Phi(\varsigma)$, and this perturbation can essentially be approached in two ways.

In the first approach, the transformation (9.10) is independent of H; this gives rise to the relations

$$\begin{aligned} \dot{\xi} &= \Psi(\xi)\,\partial_\epsilon(\xi) &&+ O(\epsilon^2) \\ H_\epsilon &= H + \epsilon\,v(H) &&+ O(\epsilon^2). \end{aligned} \tag{9.14}$$

According to equation (9.12), equation (9.13) can be averaged up to mod $O\left(\epsilon^2\right)$ with respect to the Poisson bracket in (9.7). For the quantum states to remain consistent, the averaging method has to respect the Hamiltonian structure of the quantum field theory defined in the symplectic phase space $\mathcal{A}$.

As for the second approach, one begins with a thorough description of the field v in (9.10). The condition that the mapping $\xi \mapsto \varsigma$ transforms the Poisson bracket (9.7) into (9.8) (up to mod $O\left(\epsilon^2\right)$) is encoded in the formula

$$\frac{\partial \varsigma}{\partial \xi} \Psi\left(\xi\right) \left(\frac{\partial \varsigma}{\partial \xi}\right) = \Psi\left(\varsigma\right) + \epsilon \, \Phi\left(\varsigma\right) + O\left(\epsilon^2\right). \qquad (9.15)$$

Substituting (9.10) into (9.15) yields the desired formula:

$$\Phi_{ij} = \Psi_{is} \, \partial_s v_j - \Psi_{js} \, \partial_s v_i - v_s \, \partial_s \Psi_{ij}, \qquad (9.16)$$

which implies the consistency condition

$$[\Psi, v] = \Phi. \qquad (9.17)$$

To solve Problems 1-5 means finding the global solutions to (9.17).

Let us pause to discuss the relationship between anomalies and the deformed Poisson structure. The Poisson fields given by μ on $\mathcal{A}$ (Problem 3) are all solutions of the consistency condition

$$[\Psi, \mu] = 0;$$

while the conformal fields coming out of Problem 4 are solutions of

$$[\Psi, v].$$

In these two cases, the corresponding quantum field theory is free of global anomalies. Then, depending on how the consistency conditions are handled, anomalies may still arise given that the conditions themselves assume a local character. When this is the case, the induced pathologies are purely local and, as such, they can be put into a rather manageable form using perturbation theory, mostly (9.13). Local anomalies may still manifest themselves even though the theory is shown to be global anomaly-free. For our case, this means the existence of a class of local anomalies induced by homotopic transformation. This, for instance, may appear in Lagrangian transformations in $\mathbb{R}^n$ that map Lie group induced Poisson brackets into deformed ones. The degree of accuracy needed for the counterterm to cancel the deformation is exactly what induces the local or perturbative anomaly.

Here is an illustration. Using the transformation (9.10), we define another one,

$$\Psi(\epsilon) \rightarrow \Psi(\epsilon \, \delta \epsilon).$$

The transformation $\xi \rightarrow \Xi(\xi, \epsilon)$, which transforms an initial Poisson bracket $< >_0$ to an anomalous one, namely $\{ \ \}_\epsilon$, originates from the Cauchy problem, i.e.

$$\frac{\partial}{\partial \epsilon} \Xi = v_\epsilon(\Xi), \ \ \Xi|_{\epsilon=0} = \xi.$$

Here, the vector field v_ϵ satisfies the condition (9.17), that is to say

$$[\Psi_\epsilon, v_\epsilon] = \frac{d}{d\epsilon} \Psi_\epsilon. \tag{9.18}$$

For the required homotopic transformation to be smooth, the v_ϵ solution of (9.18) ought to admit very weak singularities, so weak in fact, that the topology of $\mathcal{A}$ is not globally affected. The presence of global anomalies in this case would almost certainly imply the existence of strong singularities, and this, in turn, would violate the smoothness condition spelled out in (9.18). Do such singularities actually exist? The answer is yes. They were initially described by James Eells (circa 1960) [31], and later by Martinet [32]. We point out that even though the Eells singularities predispose to the existence of global anomalies, they do not preclude the manifestation of local anomalies. However, once again, an appropiate use of perturbation theory may put them in a manageable form, primarily by reducing their ill effects on the theory.

9.6   Nondegenerate Symplectic Spaces

In order to better grasp anomalies induced by degenerate Poisson brackets we employ the useful strategy of utilizing nondegenerate Poisson brackets as a comparison tool. Later, we will make a transition to global anomalies.

In the nondegenerate case, all points in the symplectic manifold $\mathcal{A}$, are regular, meaning that $\mathcal{A}$ is fibered by symplectic leaves Ω_i. To the symplectic leaves we associate some spaces of k-forms, $Y_k(\Omega)$, $Z_k(\Omega)$ and $H^k(\Omega)$, the later being their de Rham cohomology classes. The lack of global anomalies is due to the presence of the following smooth (i.e. non-singular) mappings

$$
\begin{aligned}
Y_k[\Omega] &= \mathbb{C}^\infty(\mathcal{A}/\Omega \rightarrow Y_k(\Omega)); \\
Z_k[\Omega] &= \mathbb{C}^\infty(\mathcal{A}/\Omega \rightarrow Z_k(\Omega)); \\
H^k[\Omega] &= \mathbb{C}^\infty(\mathcal{A}/\Omega \rightarrow H^k(\Omega)).
\end{aligned} \tag{9.19}
$$

The elements belonging to $Y_k[\Omega]$ and $Z_k[\Omega]$ are all closed forms on the symplectic leaf Ω. Thus, Ω can be described in terms of smooth coordinates on $\mathcal{A}/\Omega$ as parameters. The set of Casimir operators on $\mathcal{A}$ now reads $H^0[\Omega] \simeq C(\mathcal{A})$; as for the vector field $V(\mathcal{A})$, it preserves $C(\mathcal{A})$.

With these descriptions in hand, we proceed to extract the k-forms on a given leaf Ω. The formula

$$
\alpha(g) = \sum_{i_1 < \cdots < i_k} g_{i_1 \cdots i_k} \, \Psi_{j_1 i_1}^{-1} \, d\xi_{j_1} \wedge \cdots \wedge \Psi_{j_k i_k}^{-1} \, d\xi_{j_k}, \tag{9.20}
$$

with g a tensor, and now $\xi \in \Omega$ (instead of $\xi \in \mathbb{R}^3$ in earlier sections.) According to equation (9.20), we have the property

$$
\alpha(g \wedge g') = \alpha(g) \wedge \alpha(g').
$$

Ψ assigns to $\eta \in Y_k[\Omega]$ a k-tensor $\Psi(\eta)$. Hence, we can rederive (9.20) in terms of specific k-forms. That is,

$$
-\alpha(\psi(\eta)) = -\omega_\Omega(\Psi(\eta)) = \eta \tag{9.21}
$$

and it can be shown to preserve the original identity transformation.

The δ operator defines a Poisson mapping

$$
\delta: M^k(\mathcal{A}) \rightarrow M^{k+1}(\mathcal{A})
$$

and moreover, yields the following non-trivial condition,

$$
\delta g = [\Psi, g]. \tag{9.22}
$$

Note that M^k denotes the space of antisymmetric tensor forms of degree k. Now, using the Jacobi identity one obtains

$$[\Psi, \Psi] = 0,$$

which of course implies $\delta^2 = 0$. In geometrical terms, what we have just found is that the operator δ is a coboundary operator in the phase space $\mathcal{A}$. We will make appropiate use of this in the next few sections.

For now, we aim to discuss the conformal fields in $\mathcal{A}$, which are essential for any study of global anomalies induced by the Jacobi identity, as explained in the preceeding section.

Most Hamiltonian versions of field theories have Hamiltonian fields which are encoded in Poisson vector fields. This is most readily shown by the formula

$$\mu = \sum_{j=s+1}^{r} c_j\, z_j + \kappa, \tag{9.23}$$

where κ is the Hamiltonian field. We say of the Poisson bracket (9.7) that it is homogeneous on $\mathcal{A}$ if the symplectic form ω_Ω is zero on the cohomology class $H^2(\Omega)$. When this is the case, ω_Ω becomes

$$\omega_\Omega = \sum_{i=1}^{s} g_i\, \omega_i + d\rho_\Omega \tag{9.24}$$

with the g_i belonging to the Casimir operators $C(\mathcal{A})$, while $\rho_\Omega \in Y_{k=1}[\Omega]$. From this, we deduce that the conformal fields on $\mathcal{A}$ are given by

$$\mu = \sum_{i=1}^{s} g_i\, v_i + \Psi\, \rho_\Omega + \kappa. \tag{9.25}$$

By virtue of its nondegeneracy, the symplectic form ω generates a pairing between the tangent and cotangent space of $\mathcal{A}$. A group action $\mathcal{G}$ on the symplectic manifold $(\mathcal{A}, \omega)$ is said to be symplectic if $g \in \mathcal{G}$ defines a diffeomorphism which preserves ω, i.e.

$$g^*\omega = \omega. \tag{9.26}$$

This observation is an important ingredient in explaining the lack of global anomalies in the nondegenerate case. Diffeomorphism transformations are obviously smooth ones and so one does not expect the anomalies to arise. Actually, diffeomorphism invariance of quantum theories has been shown to prevent the occurrence of global anomalies, particularly global gravitational anomalies [15, 18].

9.7   Cancellation of Global Anomalies

The goal of this section is to answer Problems 1-5 as spelled out above. A global infinitesimal deformation of the Lagrangian $\mathcal{L}$ corresponds to a global deformation of the Poisson bracket at the quantum level. Hence, global anomalies correspond, in essence, to the formula

$$\delta\Phi = \Delta\mathcal{L}. \tag{9.27}$$

Local anomalies manifest themselves whenever the following condition holds:

$$\delta\mu = \Delta\mathcal{L}_\mu = \Phi. \tag{9.28}$$

The transformation Φ gives rise to globally trivial anomalies whenever a solution of (9.28) exists globally on $\mathcal{A}$. The Poisson bracket in (9.7) is said to be anomaly-free (global or otherwise) if each of its locally trivial transformations is globally trivial. (The case in which perturbative local anomalies arise has been treated in Section 9.3.) There is an obstruction for the Poisson bracket to be anomaly-free, and this obstruction lies in $H^2(\Omega)$.

Let us consider the case in which the bracket (9.7) is not anomaly-free. The next question of interest is to find what are the generators for the anomalies. This is especially warranted since we are aware of the obstruction keeping (9.7) from being anomaly-free. Thus, any effort aimed at determining what this obstruction looks like is welcome. A careful investigation tells us that the generators in question belong to $M^{k=1}(\mathcal{A})$ and they define closed 2-forms

$$\alpha(\Phi - \delta\omega) \in \mathbb{Z}_2[\Omega].$$

Within the framework laid out by (9.20), global anomalies can be reduced to a manageable size. In essence, this implies that the condition (9.28) is to be reduced to an equation for forms on the space $Y_{k=1}[\Omega]$. When this is achieved, one can formulate with great precision the solvability condition for (9.28) in terms of the de Rham cohomology. Here is how.

We rederive (9.28) in such a way that its corresponds to

$$d\beta = \alpha(\Phi - \delta\omega),\tag{9.29}$$

where $\beta \in Y_1[\Omega]$ is the 1-form we are looking for. Thus, solutions to (9.29) mean that the anomalies can be cancelled. Explicitly, by solving (9.29), we have

$$\mu = -\Psi^*\beta + \omega;\tag{9.30}$$

in which Ψ^* is the dual of Ψ. The right hand side of equation (9.30) is the needed counterterm to cancel the anomalies.

We now shed some light on the complex relation between anomalies and the conformal fields. In the Hamiltonian version of field theories, anomalies are omnipresent, as they often manifest themselves in the $\Phi\,dk_i$ conformal fields. These fields have a global character, as shown by the following equivalence relation:

$$\Phi\,dk_i = \Psi^*\,dh_i,\tag{9.31}$$

with $h_i \in Y_k(\mathcal{A})$. Its corresponding symplectic generator is

$$\omega = \sum_{i=1}^{r} h_i\,v_i.\tag{9.32}$$

It is a straighforward exercise to check that (9.31-32) holds in the whole phase space $\mathcal{A}$, but most particularly for simply connected Ω. However, for the non-simply connected case, there is an additional condition for ω to be well-defined in (9.32). This condition requires the cohomology class of the 1-form $\alpha(\Phi\,dk_i)$ to be trivial in $H^1[\Omega]$. Put differently, in the non-simply connected case, we have an obstruction to defining the anomaly generator.

To remove this obstruction, the first cohomology class of the symplectic leaf Ω has to vanish! (This is different from the obstruction for the Poisson bracket (9.7) to be free, since this one lies in $H^2[\Omega]$.) The h_i are determined by the set of Casimir operators. The procedure leaves ω well-defined in the degenerate case, so there is no danger of losing $\mathcal{A}'s$ symplectic structure, even in the presence of anomalies.

Now that we have elucidated the relationship between conformal fields and global anomalies, we may wonder at the anomaly constraint given by equation (9.29). Observe that the closed 2-form on the right hand side of (9.29) can be written as

$$\alpha \left(\Phi - \delta \omega \right) = \alpha \left(\Phi - \sum_{i=1}^{r} v_i \wedge \Psi^* \, dh_i \right) - \sum_{i=1}^{r} h_i \, \omega_i. \qquad (9.33)$$

This form takes its value in $H^2 [\Omega]$ and can be shown to be independent of the choice of the Hamiltonian (i.e. the h_i) in (9.31). Several facts arise in light of this finding. One is that, in the presence of global anomalies, we can still manage to preserve the consistency of the vacuum state of the quantum field theory. Adequate manipulations of (9.33) will yield a global anomaly counterterm which will cancel the manifestations of these pathologies.

The second consequence which comes to mind, and somewhat reinforces the first, is the observation that, within the appropriate constraints in the Hamiltonian version of the theory, the occurrence of global anomalies induced by deformed Poisson brackets does not consistently destroy the general consistency of the theory.

Thus, the real question before us is which of these constraints actually allows us to preserve the general consistency of the theory. According to our previous discussions, there are a handful of potential candidates. We therefore need to rephrase the question: what class of global anomalies are the least likely to destroy the consistency of the theories?

To answer this, it is useful to learn more about the anomaly Φ. Earlier, we have given a somewhat local character to Φ, whose form is given by equations (9.28-29). We said of Φ, that it is global if:

1. the cohomology classes of the 1-forms $\alpha \left(\Phi \, dk_i \right)$ are zero in $H^1 [\Omega]$ (we may recall that this particular condition defines and preserves the symplectic form ω when the leaves Ω_i are not simply-connected);

2. the quotient class of the 2-form in (9.33) is zero in $H^2 [\Omega]$.

Under condition 1, the relation (9.31) holds true and the generators for the global anomalies simply correspond to ω in (9.32). Under condition 2, we can choose the Hamiltonians h_i in (9.31) such that the primitive $\beta \in Y_{k=1} [\Omega]$ is exact in (9.33). In this case, the global solutions to equation (9.28) take the form of equation (9.30). Then, the anomaly generator in (9.32) essentially gives the non-special part of these solutions, that is, the part which has a non-zero projection on the base $\mathcal{A}/\Omega$.

We now focus on the space in which the anomalies live. Consider the following formula to be a specialized form of (9.33)

$$A = \sum_{i=1}^{r} v_i \wedge u_i \frac{1}{2} \sum_{i,j=1}^{r} b_{ij} v_i \wedge v_j. \qquad (9.34)$$

Since equation (9.34) is antisymmetric, the b_{ij} are, by definition, equivalent to $\mu_i(k_j)$ in the space of 2-cocycles. From experience, we know that the μ_i belongs to the space of 1-cocycles, consequently the b_{ij} belong to the set of Casimir operators on $\mathcal{A}$. The 3-tensor δA is a useful tool in determining the closed 3-form $\alpha(\delta A)$:

$$\alpha(\delta A) = \sum_{j=1}^{r} \omega_j \wedge \alpha \left(\mu_j - \sum_{i=1}^{r} b_{ji} v_i \right), \qquad (9.35)$$

with $\omega_j = \alpha(\delta v_j) \in Z_{k=2}[\Omega]$.

It is a relatively simple exercise to find out under what conditions equation (9.34) corresponds to a certain cocycle. Consider the relation

$$\delta A = \delta T,$$

in which T is a tensor. By applying (9.35) to this relation, we have

$$\alpha(\delta A) = \alpha(\delta T).$$

Owing to some properties of T, this reduces to

$$\alpha(\delta A) = d\alpha(T).$$

$\alpha(\delta A)$ is a *closed three-form* that is exact and well-defined in the formula

$$\alpha(\delta A) = d\beta. \qquad (9.36)$$

β is still defined as in equation (9.29). Using the formula (9.36), we verify that the 2-cocycle is indeed

$$\Phi = -\Psi\beta - A. \qquad (9.37)$$

Recall that the global anomalies are described by Φ. Thus, we have arrived at a more descriptive way of writing the anomalies. In fact, there is a more detailed form for (9.37) involving the conformal fields μ_i, namely

$$\Phi = \delta \mu + \Psi^* \mu + \sum_{i,j=1}^{r} c_{ij} z_i \wedge z_j. \qquad (9.38)$$

To obtain this result, we have proceeded as follows. First, using the formula (9.25) for 1-cocycles, we find the representation for the fields μ_i:

$$\mu_i = \sum_{j=s+1}^{r} c_{ij} z_j - \Psi^* df_i, \qquad (9.39)$$

where the c_{ij} belong to the set of Casimir operators, and $f_i \in Y_k$. The b_{ij}, introduced in formulas (9.34-35), are related to the c_{ij} by the formula

$$b_{ij} = \begin{cases} c_{ij} & \text{for } s+1 \leq j \leq r, 1 \leq i \leq r \\ 0 & \text{for } 1 \leq j \leq s, 1 \leq i \leq r. \end{cases}$$

As a consequence,

$$\alpha \left(\mu_j - \sum_{i=1}^{r} b_{ij} v_i \right) = \begin{cases} df_j & \text{for } 1 \leq i \leq r \\ -\sum_{i=1}^{r} c_{ji} \gamma_i + df_j & \text{for } 1 \leq i \leq r. \end{cases}$$

Thus, the condition (9.36) can be reduced to the form

$$\begin{aligned} d\beta &= \sum_{j=1}^{s} \omega_j \wedge df_j \\ &\quad - \sum_{j,i=s+1}^{r} c_{ji} d\gamma_j \wedge d\gamma_i \\ &\quad + \sum_{i=s+1}^{r} d\gamma_i \wedge df_i. \end{aligned} \qquad (9.40)$$

Since the 2-forms ω_i are closed, the 3-form on the left hand side of (9.40) is exact. Its primitive 2-form β is

$$\begin{aligned} \beta &= \sum_{i=1}^{s} f_i \omega_i - \tfrac{1}{2} \sum_{i,j=s+1}^{r} c_{ji} \gamma_j \wedge \gamma_i \\ &\quad + \sum_{i=s+1}^{r} \gamma_i \wedge df_i. \end{aligned} \qquad (9.41)$$

The next step is to substitute (9.39) for μ_i into (9.34). This gives

$$\begin{aligned} A &= \tfrac{1}{2} \sum_{i,j=s+1}^{r} c_{ij} v_i \wedge v_j \\ &\quad + \sum_{i,j=s+1}^{r} c_{ij} v_i \wedge \Psi^* \gamma_j \\ &\quad - \sum_{i=1}^{r} v_i \wedge \Psi^* df_i. \end{aligned} \qquad (9.42)$$

Finally, substituting (9.41) and (9.42) for β, and A into (9.37), we see that the global anomaly Φ manifests itself in equation (9.38).

For non-simply connected Ω_i, the conclusions are pretty much similar provided that we add to the right hand side of (9.38) the term

$$\left(\Psi\,\theta + \sum_{i=1}^{s} v_i \wedge \Psi^* \theta_i \right) - \sum_{i=s+1}^{r} z_j \wedge \Psi^* \eta.$$

Note that the θ_i are 1-forms satisfying

$$\sum_{i=1}^{r} \omega_i \wedge \theta_i = d\theta.$$

The Schwinger terms arise as an addition to the Poisson bracket in (9.27) and have the familiar look of the anomalies to field theorists. The situation is somewhat different in the case of global anomalies: the last summand in equation (9.38) absorbs local anomalies and thus prevents them from arising under the Schwinger terms.

9.8 Global Anomalies Induced by the Jacobi Identity

The purpose of this section is to solve Problems 4-6. The Jacobi identity (9.10) gives rise to global anomalies whenever the condition spelled out by equation (9.9) is violated. In what follows, we make explicit what is meant by global anomalies induced by the Jacobi identity. An antisymmetric 3-form is said to be an anomaly if it satisfies the condition

$$\delta B = 0, \tag{9.43}$$

and if equation (9.12) is solvable. In terms of the δ-function, the condition (9.12) takes on the form

$$\delta \Phi = B. \tag{9.44}$$

What we have done so far is to exchange the global anomaly Φ with that of B. Local anomalies, in this curent framework, are just local coboundaries of the operator δ of degree three, while global anomalies are obviously global

3-coboundaries. In general, one can expect (9.43-44) to be local anomalies whenever, at each point $k \in \mathcal{A}$, the isotropy condition

$$B\left(k_i, k_j, k_n\right) = 0 \quad (i, j, n = 1, \cdots, r) \qquad (9.45)$$

is trivial. The local character of the global anomaly (9.43-44), under the condition (9.45), is actually a rare occurrence. However, it may provide the first known example of a global anomaly degenerating into a local one, and this alone warrants further investigation of this peculiar phenomenon.

We are interested in the global solvability of equation (9.44). Our strategy will be to show that (9.44) can be reduced to (9.41) with adequate use of relations of degree less or equal to one, by which we mean forms of the type

$$(\delta A)(s) = -\delta(A(s)). \qquad (9.46)$$

For an explicit look at A, we refer the reader to equation (9.34). In order to relate (9.44) to the set of 2-forms B_i and vector fields B_{ij}, we write the relations

$$B_i = B(k_i),$$

$$B_{ij} = B(k_i, k_j);$$

with Hamitonian

$$B_{ij} - \Psi^* dh_{ij}. \qquad (9.47)$$

Furthermore, the explicit form for the 2-cocycles which are isotropic to B_i is

$$\alpha_i = \alpha \left(\sum_{j=1}^{r} \Psi^* dh_{ij} \wedge v_j - B_i\right) - \sum_{j=1}^{r} h_{ij} \omega_j. \qquad (9.48)$$

The B_i and B_{ij} are cocycles of degrees 2 and 1 respectively. The Casimir operators

$$\sigma_{ij} = \frac{(h_{ij} + h_{ji})}{2},$$

yield the conformal fields

$$h_{ij} = \sigma_{ij} + h_{ij}^0. \qquad (9.49)$$

We now work out the anomalies for the simply connected case, (i.e. the Ω_i are simply connected.) This means understanding the role played by the

2-forms α_i in equation (9.48). Applying the left and right hand sides of equation (9.44) to the k_i gives rise to a new formula for the global anomalies

$$\delta \Phi_i = -B_i, \qquad (9.50)$$

where we have used the identity $\Phi_i = \Phi(k_i)$.

Obviously, a necessary condition for the solvability of equation (9.44) is that equation (9.50) be solvable. Now the 2-forms α_i come into play: the global solvability of (9.50) corresponds to the assumption that the classes of the α_i are zero in $H^2[\Omega]$. On the basis of this, we can choose the h_{ij} from (9.47) in such a way that the σ_{ij} and h^0_{ij} have representation as in equation (9.49). When this is achieved, the form α_i can be written as

$$\alpha_i = d\beta_i, \qquad (9.51)$$

or equivalently,

$$\alpha_i = \alpha_i^0 \sum_{J=s+1}^r h^0_{iJ} d\gamma_J - \sum_{J=s+1}^r \sigma_{iJ} d\gamma_J \\ - \sum_{j=1}^s h^0_{ij} \omega_j - \sum_{j=1}^s \sigma_{ij} \omega_j. \qquad (9.52)$$

Notice that α_i^0 corresponds to the first summand in equation (9.46); these 2-forms are independent of any choice of h_{ij}.

On the symplectic leaf Ω we choose basic 2-cycles $K_1, \cdots, K_n$ which can be thought of as adjoint to the 2-forms $\omega_1, \cdots \omega_j$:

$$\int_{K_i} \omega_j = \delta_{ij}$$

Hence, according to (9.52), the condition (9.51) can be reformulated as

$$\sigma_{ij} = \int_{K_j} \left(\alpha_i^0 - \sum_{J=1}^r h^0_{iJ} d\gamma_J - \sum_{l=1}^s h^0_{il} \omega_l \right) = 0. \qquad (9.53)$$

One consequence of equations (9.47) and (9.49) is the freedom they give in adding arbitrary operators, particularly those of the c_{ij} type. Recall that these belong to $C(\mathcal{A})$, the set of Casimir operators on the phase space $\mathcal{A}$. In (9.52), the σ_{ij} could be replaced by $\sigma' = \sigma_{ij} - c_{ij}$: the σ_{ij} are fixed by (9.52) while the c_{ij} are arbitrary. This allows one to write a uniquely determined relation:

$$\sum_{i,j=1}^s \sigma_{ij} [\omega_i \wedge \omega_j] = 0. \qquad (9.54)$$

Later, we shall make appropiate use of this.

When the global anomaly B satisfies (9.51), it has a topological defect in the form of an $s \times s$ matrix $\sigma = \sigma_{ij}$ [2]. This defect is detectable under the anomaly condition in (9.51) using the formula

$$\Phi_i = -\Psi^* \beta_i + \sum_{j=1}^{r} h_{ij} v_j, \qquad (9.55)$$

which, in some ways, is a globalization of (9.30).

Our next order of business is to investigate the behavior of the global anomaly (9.43-44) in $\mathcal{A}$. As we have seen in previous sections, the procedural approach will be to reduce (9.44) to an equation of forms using, among other techniques

$$\delta \omega (k_i) = -B_i,$$

($B - \delta \omega$ in this case). Recall that ω has the dual role of being a symplectic 2-form on $\mathcal{A}$, and also a generator of anomaly (the most vivid illustration of which is contained in equation (9.32)). Using the equivalence relation

$$\omega (k_i) = \Phi_i$$

we deduce that ω can be reconstructed from Φ_i in (9.55). There is a hitch however, namely, that the h_{ij} be antisymmetric.

For the global anomaly B to exist, the topological defect, σ, must be trivial. When this is the case, the antisymmetric property of the h_{ij} allows us to choose the Hamiltonian in a such a way that the α_i are exact. Thus, the explicit form of the anomaly generator ω is

$$\omega = -\sum_{i=1}^{r} v_i \wedge \Psi^* B_i + \frac{1}{2} \sum_{i,j=1}^{r} h_{ij} v_i \wedge v_j. \qquad (9.56)$$

As a consequence, the 3-form

$$\begin{aligned}
\alpha (B - \delta \omega) &= \alpha (B_i) \\
&\quad - \sum_{i=1}^{r} v_i \wedge B_i \\
&\quad - \tfrac{1}{2} \sum_{i,j=1}^{r} (v_i \wedge v_j \wedge \Psi^* dh_{ij}) \\
&\quad - \sum_{i=1}^{r} \omega_i \wedge B_i,
\end{aligned} \qquad (9.57)$$

corresponds to the global solvability of equation (9.44).

We now list the conditions under which the cocycle B is a global anomaly.

1. First, B must satisfy local anomalies condition;

2. The α_i in (9.52) ought to be trivial in $H^2[\Omega]$;

3. The topological defect σ_{ij} of B also ought to be trivial;

4. The 3-form α_i given in (9.48) must be exact, i.e. $\alpha(B - \delta\omega) = d\beta_B$.

When these conditions are all satisfied, we obtain the formula for global Poisson bracket deformation:

$$\Phi = -\Psi^* \beta_B + \sum_{i=1}^{r} v_i \wedge \left(-\Psi^* \beta_i + \frac{1}{2} \sum_{j=1}^{r} h_{ij}\, v_j \right), \qquad (9.58)$$

where the 1-forms β_i were borrowed from equation (9.51). If we consider the B_{ij} as being strictly Hamiltonian, then the conditions (1) to (4) apply to the case of non-simply connected Ω_i.

In the space of 3-forms on $\mathcal{A}$, the global anomaly is given by the formula

$$\begin{aligned}
B &= \delta M + \Psi^* \mu + \sum_{i,j,k=s+1}^{s} c_{ijk}\, z_i \wedge z_j \wedge z_k \\
&\quad + \sum_{i,j=s+1}^{r} \Psi^*(\nu_{ij})\, z_i \wedge z_j \\
&\quad - \left(\Psi^* \tau + \sum_{i=1}^{4} v_i \wedge \Psi^* \tau_i - \frac{1}{2} \sum_{i,j=1}^{s} v_i \wedge v_j \wedge \Psi^* \tau_{ij} \right) \\
&\quad - \left(\sum_{i=s+1}^{r} z_i \wedge \Psi^* O_i + \sum_{i=1}^{s} \sum_{j=s+1}^{r} v_i \wedge z_j \wedge \Psi^* O_{ij} \right).
\end{aligned} \qquad (9.59)$$

A few definitions are in order. M belongs to the space of 2-forms on $\mathcal{A}$, ν_{ij} are elements of $Z_{k=1}[\Omega]$, $\tau \in Y_{k=2}[\Omega]$, the $\tau_i \in Y_{k=2}[\Omega]$, while $O_{ij} \in Y_{k=1}[\Omega]$. Their explicit forms are revealed by

$$\begin{aligned}
d\tau_{ij} &= 0 \\
d\tau_i &= \sum_{j=1}^{s} \omega_j \wedge \tau_{ij} \\
d\tau &= \sum_{i=1}^{s} \omega_i \wedge \tau_i,
\end{aligned} \qquad (9.60)$$

and

$$\begin{aligned}
dO_{ij} &= 0 \\
dO_i &= \sum_{i=1}^{s} \omega_j \wedge O_{ij}.
\end{aligned} \qquad (9.61)$$

9.9   Generalized Wess-Zumino Consistency Condition

Our primary goal here is to compute the cocycles and coboundaries of the δ operator:

$$\delta : M^{q,p+1} \rightarrow M^{q+2,p}.$$

It acts by the formula

$$(\delta \eta)_{i_1 \cdots i_p} = \sum_{j=1}^{r} \omega_j \wedge \eta_{i_1 \cdots i_p j},$$

where the ω_i are essentially the same as before, and $M^{q,p}$ denotes the space of p-forms on Ω.

We have the condition

$$\sum_{l \leq i \leq s} [\eta_i \wedge \omega_i] = 0, \tag{9.62}$$

from which we define the δ on the direct sum

$$M_k = \bigoplus_{p+q=k} M^{q,p}, \delta : M_k \rightarrow M_{k+1}.$$

Using (9.62) gives

$$\delta^2 = 0, \ d\delta - \delta d = 0.$$

With these descriptions in hand, we can proceed to write a generalized version of the Wess-Zumino consistency condition [33]. The Wess-Zumino consistency condition represents a simple test of anomalies detection. In our context, this condition corresponds to the following linear chain of obstructions which are essentially towers of obstructions in double complexes:

$$
\begin{aligned}
d\eta_{0,k} &= 0 \\
d\eta_{l,k} &= \omega_j \wedge \eta_0 \\
d\eta_{k-p} &= \omega_j \wedge \eta_{k-p} \\
d\eta_{k-l} &= \omega_j \wedge \eta_{k-2} \\
d\eta_k &= \omega_j \wedge \eta_{k-l}.
\end{aligned}
\tag{9.63}
$$

9.10   References

[1] Baadhio, R. A.: *Global Anomalies Induced by Deformed Quantization*, University of California, Berkeley Preprint UCB-PTH-94/20 and LBL-35952, December 1995. Submitted for publication.

[2] Dirac, P. A. M.: **The Principles of Quantum Mechanics**, Clarenden Press, Oxford 1930.

[3] Weinstein, A.: **Lectures on Symplectic Manifolds**, Conf. Board Math. Sciences Reg. Conf. Ser. Math. Vol. 29, American Math. Society, 1979 Providence, R. I.

[4] Kostant B.: Lectures Notes Mathematics 170 (1970) 87.

– Sympo. Mathematics 14 (1974) 139.

[5] Bayen, F., Flato, M., Fronsdal, C., Lichnerowicz, A., and Sternheimer, D.: *Deformation Theory and Quantization. I. Deformations of Symplectic Structures*, Ann. Physics 111 (1978) 61-110.

– (Idem): *Deformation Theory and Quantization. II. Physical Applications*, Ann. Physics 111 (1978) 111-151.

[6] Moyal, J.: *Quantum Mechanics as a Statistical Theory*, Proc. Cambridge Phyl. Society 120 (1949) 99-124.

[7] Weyl, H.: **The Theory of Groups and Quantum Mechanics**, Dover (1931) New York.

[8] Lie, S.: **Theorie der Transformationsgruppen**, Leipzig, Teubner 1890.

[9] Weinstein, A.: *Deformation Quantization*, Seminaire BOURBAKI 789, 46 Years 1993-1994. To appear.

[10] Gutt, S.: *Equivalence of Twisted Products on a Symplectic Manifold*, Letters Math. Physics 3 (1979) 495-502.

[11] Vey, J.: *Deformation du crochet de Poisson sur une varieté symplectique*, Comment. Math. Helvetica 50 (1975), 421-454.

[12] De Wilde, M. and Lecomte, P.: *Formal Deformations of the Poisson Lie Algebra of a Symplectic Manifold and Star Products. Existence, Equivalence, Derivations*, in **Deformation Theory of Algebra and Structures**

and Applications. Eds. Hazewinkel, M. and Gerstenhaber, M. Kluwer Academic Publications (1988) 897-960.

[13] Witten, E.: *Topological Tools in* 10-*Dimensional Physics*, Int. Journal Modern Physics A, vol 1 no. 1 (1986) 37-64.

[14] Witten, E.: *An* SU (2) *Anomaly*, Physics Letters B117 no. 5 (1982) 324-328.

— *Global Gravitational Anomalies*, Commun. Math. Physics 100 (1985) 197-229.

[15] Baadhio, R. A.: *Global Gravitational Anomaly-Free Topological Field Theory*, Phys. Letters B299 (1993) 37-40.

[16] Baadhio, R. A.: *Knot Theory, Exotic Spheres and Global Gravitational Anomalies*, in **Quantum Topology**, 78-90, by L. H. Kauffman and R. A. Baadhio, World Scientific, 1993 Singapore, London, New Jersey.

[17] Baadhio, R. A. and Kauffman, L. H.: *Link Manifolds and Global Gravitational Anomalies*, Rev. Math. Physics 5 N.2 (1993) 331-343.

[18] Baadhio, R. A.: *Mapping Class Groups for* $D = 2 + 1$ *Quantum Gravity and Topological Quantum Field Theories*, Nuclear Physics B441 Nos. 1-2 (1995) 383-401.

[19] See reference [3].

[20] Arnold, V. I. and Givental' A. B.: **Symplectic Geometry**, Springer-Verlag, 1990 Berlin, New York.

[21] Witten, E.: *Quantum Field Theory and the Jones Polynomial*, Comm. Math. Physics 121 (1989) 351-399.

[22] Schaller, P. and Schwarz, G.: *Anomalies From Geometric Quantization of Fermionic Fields*, J. Math. Physics 31 N. 10 (1990) 2366-2377.

[23] Morita, K. and Kaso, H.: *Anomalies From Stochastic Quantization and their Path Integral Interpretation*, Phys. Reviews D41 N2 (1990) 553-560.

[24] Morita, K.: *Note on the Anomalies From Schotastic Quantization*, Prog. Theor. Physics 81 N6 (1989) 1099-1103.

[25] Panfil, S. L.: *Canonical Quantization, the Dirac Problem and the Anomalies*, Mod. Phys. Letters A4 N26 (1989) 2561-2167.

[26] Krammer, U. and Rebhan A.: *Anomalous Anomalies in the Carlip-Kallosh Quantization of the Green-Schwarz Superstring*, Phys. Letters B236 N3 (1990) 255-261.

[27] Kim, Y. B.: *Chiral Anomalies of Higher Derivative Theories from Stochastic Quantization*, Mod. Phys. Lett. A7 N30 (1992) 2861-2866.

[28] Van Doren, S. and Van Proeyen, A.: *Simplification in Lagrangian BV Quantization Exemplified by the Anomalies of Chiral W_3 Gravity*, Nucl. Physics B411 N1 (1994) 257-306.

[29] Eells, J.: **Singularities of Smooth Maps**, Columbia University Mathematics Publication CU-5-ONR-266 (57)-M 1960 New York.

[30] Martinet, J.: **Singularities of Smooth Functions and Maps**, London Math. Society Lecture Note Series 58, Cambridge University Press, 1982 Cambridge, New York.

[31] Wess, J. and Zumino, B.: *Consequences of the Anomalous Ward Identities*, Phys. Letters B73 (1971) 95.

[32] Axelrod, S., Della Pietra, S. D. and Witten, E.: *Geometric Quantization of Chern-Simons Gauge Theory*, J. Diff. Geometry 33 (1991) 787-902.

[33] Atiyah, M. F.: **The Geometry and Physics of Knots**, Cambridge University Press, 1990, Cambridge, New York.

Chapter 10

Chiral and Gravitational Anomalies

Whenever symmetries of a classical theory are broken by quantum corrections, anomalies are bound to arise. Anomalies originating from *global* gauge or diffeormorphic (i.e. non-identity component general coordinate) transformations are referred to as global gauge or global gravitational anomalies. This class of pathologies is far more subtle and difficult to evaluate, especially in higher dimensions, than the ones we are about to introduce: chiral and gravitational anomalies. We will dedicate more than one chapter to global anomalies, but before we do so, it is essential that we give some background on the subject. This is the goal of this chapter. The upcoming presentation is self-contained.

A celebrated example in the study of classical anomalies (as opposed to global anomalies) is that of the so-called triangle diagram endowed with two vector currents and an axial vector current. The lack of conservation of the axial vector is manifest whenever Bose symmetry and vector current conservation are required. As a consequence, a breakdown of chiral symmetry, most notably in the presence of external gauge fields, is exhibited. Therefore, even though requiring Bose symmetry and vector current conservation may be important for internal consistency reasons, the procedure gives rise to anomalies and challenges the theory's overall consistency. The existence of such anomalies has led to an understanding of π° decay, and later, to the resolution of the $U(1)$-problem in Quantum Chromodynamics (QCD) [1]. Anomalies of this type essentially represent the breakdown of global axial symmetries; their existence in no way jeopardizes unitarity or renormalizability. In contrast, anomalies which arise when chiral currents are coupled to gauge fields are of a more serious nature since theories which are not gauge invariant suffer from nonrenormalizability.

In what follows, we shall consider non-abelian gauge theories coupled to fermions in some representations R^a of the gauge group G to which Feynman's rules apply. Anomalies for gauge currents in this class of theories are fairly well known: they are proportional to a purely group-theoretic factor times a given Feynman integral. To get rid of the anomaly, the sum of the group theoretic factor over various representations ought to vanish, as is readily shown by the formula:

$$I = \sum_{R_L} \mathrm{S\,Tr}\, R_L^a R_L^b R_L^c - \sum_{R_R} \mathrm{S\,Tr}\, R_R^a R_R^b R_R^c = 0. \qquad (10.1)$$

Here, R_L (repectively R_R) stands for the representations of G carried by the left (respectively right) handed fermions; S Tr denotes the symmetrized trace over the group generators involved. Formula (10.1) has a simple yet far-reaching consequence, namely, any given theory will not be gauge-invariant whenever (10.1) is not trivial. This fact is well demonstrated in dimension four where, for instance, the remarkable cancellation of anomalies within the framework of the Glashow-Salam-Weinberg theory is a direct consequence of the vanishing of (10.1). The result arises by writing the known quarks and leptons in terms of left-handed Weyl fermions. It then follows that for each family (there are three of them), we can write down SU(3) × SU(2) × U(1) quantum numbers as

$$(3,2)_{1/2} \oplus (1,2)_{-1/2} \oplus (\overline{3},1)_{1/3} \oplus (1,1)_1. \qquad (10.2)$$

Several facts about formula (10.2) are of interest. Firstly, the $U(1)$ hypercharge anomaly cancels because of the contribution of every member of the family. Secondly, it is a source of motivation for understanding the quantum numbers of the quarks and leptons, thus incidentally renewing our interest in grand unified [2] and Kaluza-Klein theories [3].

Theories that allow vector currents to be replaced by the energy-momentum tensor are theories in which the axial vector symmetry is violated in the presence of external gravitational fields. In order to illustrate this, we pick a $U(1)$ gauge field coupled to some $V - A$ currents with external gravitons. We then have a tensor proportional to the trace of K, the generator of $U(1)$. The case $\text{Tr}\, K = 0$ is of importance given that it is impossible to simultaneously maintain the conservation of the energy-momentum tensor and the $U(1)$ gauge symmetry, as shown in great detail by Alvarez-Gaumé and Witten [4]. Therefore, to cancel the anomaly requires that $\text{Tr}\, K$ vanish for the generator of a given $U(1)$ factor in the gauge group G. The quantum numbers in (10.2) do indeed satisfy this consistency condition.

As a general rule, $\text{Tr}\, K$ can be shown to vanish whenever the theory's low energy group originates from a bigger, simple or semi-simple unified gauge group. The reason for this is that the trace of any generator *always* vanishes for a compact, simple group. The standard $\text{SU}(3) \times \text{SU}(2) \times U(1)$ theory to which we alluded earlier is among the class of theories known to satisfy the general rule on the vanishing of $\text{Tr}\, K$, and it applies to each of the three families.

We now focus on anomalies in higher dimensions. In dimensions $2n$, chiral fermions coupled to gauge fields have localized anomalies because of the requirement of Bose symmetry for external gauge lines. Furthermore, each vertex contains a factor R^a, and this alone implies that the group theory factor appearing in the anomalies can easily be extracted; explicitly:

$$I_{\text{Tr}} = \sum_{R_L} \text{S Tr}\, (R_L^{a_1} \cdots R_L^{a_{n+1}}) - \sum_{R_R} \text{S Tr}\, (R_R^{a_1} \cdots R_R^{a_{n+1}}). \qquad (10.3)$$

Cancelling (10.3) when $n > 2$ is far more complex than cancelling (10.1). It is therefore of no surprise that the set of anomaly-free chiral representations of fermions is considerably reduced in dimension four. Accordingly, it is only fair to say that anomalies in general restrict the number of potential consistent theories in four dimensions, but even more so in higher dimensions. In dimensions $4k + 2$, for instance, the energy-momentum tensor for chiral fermions is anomalous. Alvarez-Gaumé and Witten [4] found that theories defined in even dimensions often suffer from anomalies, localized in gauge currents (the so-called gauge anomalies), and in the energy-momentum tensor (i.e. anomalies of a gravitational nature). Additional anomalies are mixed anomalies that correspond to graphs containing gravitons and gauge fields. A theory is said to be consistent when all the anomalies have been cancelled.

This leaves us with a very small class of theories that are chiral and anomaly-free in even dimensions.

When taking supersymmetry into account [21] there are, to date, only two types of anomaly-free theories: $N = 2$ chiral supergravity theory in ten dimensions and $N = 1$ super Yang-Mills theory coupled to $N = 1$ supergravity, also in dimension ten. The latter has $E_8 \times E_8$ or $SO(32)$ gauge group and the theory is called the heterotic superstring theory [5, 6]. The anomaly cancellation coupled to the real possibility that this class of theories provides an ultraviolet finite quantum theory of gravity are perhaps the driving forces that sustain our interest in string theory.

Physicists working in the field of anomalies have several tools available today in furthering their investigations. Among them is index theory [7] and related applications, exhibiting in the process the ever-increasing interaction between mathematics and physics. The interaction in question began in the mid-seventies with the discovery by Jackiw et al. [8] that the global $U(1)$ axial anomaly could be understood in terms of the Atiyah-Singer Index theorem. The basis for this was the realization that the expectation value of the divergence of the axial current can be related to the spectral asymmetry of the Dirac operator. The authors, in reference [8], demonstrated that the spectral asymmetry is indeed concentrated in the space of zero modes, some of whose properties are determined by index theorem arguments. Independently of obtaining the correct form and normalization of the $U(1)$ anomaly using index theory, the approach provides a powerful way of analyzing some subtleties about the dynamics of fermionic gauge fields.

Later, even more powerful forms of the index theorem, namely index theorem for families of elliptic operators [9, 10], were used to provide an understanding of the non-abelian anomaly as well as that of gravitational anomalies [11, 12, 13]. These results were obtained within the Euclidean formulation of field theory. Yet, it is now also possible to understand them from an algebraic [14, 15] or topological [16] point of view in the Hamiltonian formalism. The study of global gauge [17] and gravitational [18] anomalies requires the use of spectral flows and index theory for families of operators and a good deal of algebraic topology, as we shall see in upcoming chapters.

10.1 Spinor Representations of Lorentz Groups

Chiral fermions are among the basic ingredients in the study of anomalies, and this alone warrants the present review on spinor representations. To avoid infrared problems we will work on Euclidean space compactified to a sphere. This review is not exhaustive. The interested reader is referred to Cartan's book **Theory of Spinors**, Dover, New York, 1981 for a detailed study of spinors. We will be concerned primarily with spinor representations of Lorentz groups in Euclidean and Minkowski spaces. Our starting point is the Clifford algebra in dimensions $2n$. A flat metric g^{kl} with signature (t, s) reads:

$$\{\Gamma^k, \Gamma^l\} = 2g^{kl}; \tag{10.4}$$

$k, l = 0, \cdots, 2n - 1$. A Dirac spinor is a field which, under an infinitesimal $SO(t, s)$ transformation changes as

$$\begin{aligned}
\delta\Psi &= -\tfrac{1}{2}\, \epsilon_{kl}\, \Sigma^{kl}\, \Psi \\
\Sigma^{kl} &= -\tfrac{1}{4}\, \left[\Gamma^k, \Gamma^l\right] \\
\Gamma^{+\dagger} &= \Gamma^1, \cdots, \Gamma^{t\dagger}.
\end{aligned} \tag{10.5}$$

In even dimensions, $\bar{\Gamma}$ is a matrix which anticommutes with Γ^k:

$$\bar{\Gamma} = \alpha\, \Gamma^0\, \Gamma^1 \cdots \Gamma^{2n-1},$$

where we think of α as a phase chosen in such a way that $\bar{\Gamma}^2 = 1$, that is to say

$$\begin{aligned}
\alpha^2 &= (-1)^{(s-t)/2}; \\
\alpha^\star &= (-1)^{(s-t)/2}\, \alpha.
\end{aligned}$$

Notice that $\{\bar{\Gamma}, \Gamma^k\}$ is trivial and so $\bar{\Gamma}$ is free to commute with the generators of the Lorentz group.

Using $\bar{\Gamma}$, we define two projection operators:

$$\begin{aligned}
\bar{p}_\pm &= \frac{1}{2}\left(1 \pm \bar{\Gamma}\right), \\
\Psi_\pm &= p_\pm\, \Psi.
\end{aligned}$$

We say of Ψ_+ (resp. Ψ_-) that they are positive (resp. negative) chiral Weyl spinors, and it follows that

$$\begin{aligned}
\bar{\Gamma}\,\Psi_+ &= \Psi_+; \\
\bar{\Gamma}\,\Psi_- &= -\Psi_-.
\end{aligned} \tag{10.6}$$

The Clifford algebra in (10.4) has a unique faithful representation of dimension $2n$. Hence $\Gamma^k, \Gamma^{k*}, -\Gamma^{k\dagger}$ are all related owing to the fact that they satisfy the same anticommutation relation (10.4). Next, we need to find a charge conjugate Dirac spinor. To this end we write down the matrix M as

$$\left(\Sigma^{kl}\right)^* = M\,\Sigma^{kl}\,M^{-1}, \tag{10.7}$$

which gives

$$\Psi^c = c\Psi \equiv M^{-1}\,\Psi^*. \tag{10.8}$$

From (10.6) we deduce that Ψ^c and Ψ have identical Lorentz transformation properties, which are:

$$\delta\Psi^c = -\tfrac{1}{2}\,\epsilon_{kl}\,\Sigma^{kl}\,\Psi^c$$
$$\left[c, \Sigma^{kl}\right] = 0. \tag{10.9}$$

When $c^2 = +1$, the matrix M gives rise to projection operators $(1 \pm c)/2$ and to the following Majorana and anti-Majorana fields:

$$\Psi_k = (1 + c)\,\frac{\Psi}{2}$$
$$\Psi_{\bar{k}} = (1 - c)\,\frac{\Psi}{2}. \tag{10.10}$$

A question of interest is whether c and $\bar{\Gamma}$ commute. To answer, we begin by writing $\bar{\Gamma}$ in terms of the Σ^{kl}; this is readily done by the formula:

$$\bar{\Gamma} = (-2)^n\,\alpha\,\Sigma^{01}\,\Sigma^{23}\,\cdots\,\Sigma^{2n-2,\,2n-1}, \tag{10.11}$$

which yields

$$c\,\bar{\Gamma} = (-1)^{(s-t)/2}\,\bar{\Gamma}\,c. \tag{10.12}$$

Let us investigate the case in which $(s - t)/2$ is odd. According to reference [4] c flips the chirality of fermions. However, for $(s - t)/2$ even, c does not flip the chirality. In dimensions $4k$, $\{c, \bar{\Gamma}\} = 0$ whereas in dimensions $4k + 2$, $\left[c, \bar{\Gamma}\right] = 0$. In order to extract the physical importance behind these results, one considers the solutions to the massless Dirac equation in dimension $2k$. A free, massless fermion with definite momentum moving in $2k - 1$ dimensions possesses a space wave momentum function which is none other than

$$\Gamma^0\,u(p) = \Gamma^{2k-1}\,u(p). \tag{10.13}$$

The helicity operator reads

$$h = \Sigma^{12}\,\Sigma^{34} \cdots \Sigma^{2k-3,\,2k-2} \sim \Gamma^1\,\Gamma^2 \cdots \Gamma^{2k-2}. \tag{10.14}$$

Combining (10.12) to (10.13) gives

$$\begin{aligned}
h\,u\,(p) &= h\,\Gamma^0\,\Gamma^0\,u\,(p) \\
&= \Gamma^0\,h\,\Gamma^{2k-1}\,u\,(p) \\
&= \bar{\Gamma}\,u\,(p).
\end{aligned} \tag{10.15}$$

Formula (10.15) states the well known fact that helicity and chirality are unique manifestations of one another in dimension four. Accordingly, in dimensions $4k$, Minkowskian space charge conjugation flips helicity, whereas this does not hold true in $4k + 2$ dimensions.

Euclidean $4k$-dimensional spinors are similar in several ways to $(4k+2)$-dimensional Minkowskian spinors, and to $4k+2$ dimensional spinors like the one in $4k$ dimensional Minkowskian space. Anomaly wise, an interesting fact about dimension $4k + 2$ is the triviality of the term $\left[c, \bar{\Gamma}\right]$, and the consequent imposition of the Weyl-Majorana condition. To impose, for example, a simultaneous Weyl-Majorana condition would require dimension $2 \bmod 8$. Results of this type are important if one is to build minimal on-shell multiplet representations of higher dimensional supersymmetry and supergravity theories. Massless vector fields in dimension ten, for instance, represent eight degrees of freedom - the massless state group in dimension $D = 10$ is SO (8) - and the state created by a ten-dimensional vector field transforms like a vector under SO (8).

To comment further on the simultaneous imposition of the Weyl-Majorana condition, consider a supersymmetric multiplet together with a vector field. Imposing the condition in question requires first finding a fermionic field whose degree of freedom is similar to that of the multiplet. In dimension ten, any given Weyl fermion possesses exactly $2^4 = 16$ degrees of freedom. Thus, imposing both the Majorana and Weyl conditions reduces the number of degrees of freedom to eight. As we have explained, such conditions are necessary for the overall internal consistency of the theory. Anomalies in this case often act to substantially reduce the theory's number of degrees of freedom.

10.2   The Group Invariant I_{Tr}

The evaluation of $\text{S Tr}\, R_L \cdots R_R$ is essential for any thorough analysis of the anomalies. To evaluate this term amounts to computing the quantity $\text{Tr}_r\, N^n$, where N is a matrix in the representation r of the Lie algebra of interest. It is important to differentiate the cases in which n is even or odd. For even $n = 2k$, $\text{Tr}\, N^{2k}$ is positive definite; thus, a theory with only left-handed fermions in these dimensions will always be anomalous. A celebrated such instance is the $N = 1$ super Yang-Mills theory in ten dimensions. The theory is indeed anomalous for any gauge group G [4]. For odd $n = 2k+1$, we have a purely left-handed anomaly-free representation. A detailed analysis of groups and their representations is often necessary when investigating anomalies. Classically, one works with either of the following representations:

(i) the real symmetric (also called simply real) representation in which the group matrices in the representation are equivalent to their complex conjugates. Also noted in this case is the possibility to define a basis in which the Hermitian generators of the representation are purely imaginary;

(ii) the real antisymmetric (or pseudo-real) representation. This case is somewhat similar to (i) except that there is no basis in which the Hermitian generators can be chosen as purely imaginary matrices. A readily available example is the two-dimensional representation of SU (2). The Hermitian generators are the Pauli matrices:

$$\sigma_i/2, \quad i = 1, 2, 3$$
$$c = \sigma^2$$
$$\sigma_i^\dagger = -\sigma^2 \, \sigma^i \, \sigma^2$$
$$c^\dagger = -c.$$

In these, there is no possible basis in which the σ_i's could be defined as purely imaginary.

(iii) the complex representation. In contrast to (i), we encounter complex representations whenever the group matrices in the representation are not equivalent to their complex conjugates.

We use these cases to illustrate that $\text{Tr}\, N^{2k+1}$ vanishes for real or pseudoreal representations, i.e.

$$
\begin{aligned}
\text{Tr}\, N^{2k+1} &= \text{Tr}\left(N^{2k+1}\right)^{\dagger} \\
&= \text{Tr}\left(N^{\dagger}\right)^{2k+1} \\
&= -\text{Tr}\left(c^{-1} N c\right)^{2k+1} \\
&= 0.
\end{aligned}
\tag{10.16}
$$

Next, we focus on some given groups and their representations in dimensions $4k$. We begin with $\text{SO}\,(2k+1)$. This group has two basic representations: a vector representation which is obviously real, and a spinor representation. The later is unique. Moreover, it is equivalent to its complex conjugate. This leaves open the possibility that the $\text{SO}\,(2k+1)$ spinor representation is either real or pseudoreal. Incidentally, it is worth noticing that all representations of $\text{SO}\,(2k+1)$ can be obtained using suitable tensor products of these two representations. Therefore, all representations of $\text{SO}\,(2k+1)$ are either real or pseudoreal. Hence, a theory whose gauge group is $\text{SO}\,(2k+1)$ is anomaly-free.

Similar conclusions apply to the symplectic group $\text{Sp}(n)$, and the exceptional groups G_2, F_4, E_7, and E_8. The fundamental representations of G_2, F_4, E_7, and E_8 have respective dimensions 7 (real), 26 (pseudoreal), 56 (pseudoreal), 248 (real); all other representations are either real or pseudoreal. Potentially anomalous candidates are $\text{SO}\,(2k)$, $\text{SU}\,(k)$, and E_6. Our previous discussion of spinors in Euclidean space tells us that $\text{SO}\,(4k)$ spinors are not complex, and this leaves $\text{SO}\,(4k)$ anomaly-free. Concerning $\text{SO}\,(4k+2)$, the situation is far from being safe. For one thing, $\text{SO}\,(4k+2)$ complex conjugation exchanges the spinor representations, thus indicating the complex character of their nature. As for $\text{SU}\,(k)$, for different values of k, they suffer from anomalies in virtually all dimensions. E_6 is an exceptional group with 78 generators and complex fundamental representations 27, $\overline{27}$. There is an anomaly in the 27 representation which cancels in dimension four [5, 6], but it does not in dimensions $4k$, whenever $k \geq 2$. One works out the E_6 anomaly by embedding E_6 in

$$
E_6 \supset \text{SO}\,(10) \times \text{U}\,(1).
$$

The 27 representation now reads

$$
27 = 1(4) + 10(-2) + 16(1);
$$

and the anomaly is now computed with respect to $SO(10) \times U(1)$.

10.3   The Green-Schwarz Anomaly Cancellation

In this section, we will discuss the Green-Schwarz anomaly cancellation for the $E_8 \times E_8$ or $SO(32)$, the heterotic superstring theory [5, 6]. Consider a Lie algebra $\mathcal{G}$ whose maximal subset of commuting generators (the so-called Cartan subalgebra) is B_i, $i = 1, \cdots, r$; where r is the rank of $\mathcal{G}$. The eigenvalues of the vector $\vec{B}$ on the states of the representation $\vec{B}|\vec{\alpha}> = \vec{\alpha}|\vec{\alpha}>$ labeled all possible representations. We think of $\vec{\alpha}$ as a weight vector with r components. The formula

$$\text{Tr}\, K(g) = \sum_{\vec{\alpha}} \exp i\vec{x} \cdot \vec{\alpha}, \qquad (10.17)$$

gives the character of g in the representation K, where K is a group matrix. Applying the observation that any group element K can be represented as $\exp N$, one deduces that the computation of equation (10.17) reduces to computing all the traces of the form $\text{Tr}\, N^n$. Weyl, in reference [19], has provided us with a general formula aimed at computing the form (10.17). Unfortunately, as this formula stands, it has little relevance to the anomaly cancellation. There is, however, a fruitful approach to the computation of the character (10.17). Essentially, the procedure involves computing (10.17) in terms of the character of the fundamental representation of $\mathcal{G}$. For some representations R, $\text{Tr}\, N^n$ can be written in terms of traces and products of traces in the fundamental representations. Following Green and Schwarz's notations in [5, 20], we denote traces in the fundamental representation by lower case, i.e. $\text{tr}\, N^n$:

$$\text{Tr}\, N^n = e_1\, \text{tr} N^n + e_2\, \text{tr} N^2\, \text{tr} N^{n-2} + \cdots \qquad (10.18)$$

The number of independent coefficients e_i needed to be taken into account in the analysis of anomaly cancellation thus depends on the number of irreducible traces $\text{tr} N^n$; these are traces that cannot be factorized into products of lower-order traces. In fact, this number depends on the Casimir operators of the group $\mathcal{G}$.

Group theoretic facts show that the number of independent traces for a simple Lie group is given by the rank of G. For instance, SU(3) enjoys two independent Casimir operators which are related to $\text{tr}(N^2)$ and $\text{tr}(N^3)$. For simple enough representations, one can calculate the coefficients of (10.18) without much trouble.

We are interested in computing $\text{Tr}\,(N^n)$ for the adjoint representation of SO(n) or Sp(n) mainly to illustrate the discussion above. The standard prescription for doing so is to first build the adjoint representation from tensor products of the fundamental representation. In the fundamental representation of SO(n) or Sp(n) lives a generic group, call it Q_{ij}, whose action on the representation space reads:

$$t_i \rightarrow Q_{ij}\,t_j.$$

For SO(n), the adjoint representation is equivalent to the antisymmetric tensor irreducible representation. We obtain as a consequence of this a new transformation,

$$t_{ij} \rightarrow \frac{1}{2}\left(Q_{ik}\,Q_{jl} - Q_{il}\,Q_{jk}\right) t_{kl}. \qquad (10.19)$$

In the symplectic case, the group Sp(n) has an invariant, an antisymmetric bilinear form which has to be taken into account. The adjoint representation that follows from the symmetric second-rank tensor is:

$$t_{ij} \rightarrow \frac{1}{2}\left(Q_{ik}\,Q_{jl} + Q_{il}\,Q_{jk}\right) t_{kl}. \qquad (10.20)$$

We are now in a good position to determine the character of the adjoint representation in terms of the character of the fundamental irreducible representation. It is encoded in the formula:

$$\text{Tr}\,Q = \frac{1}{2}\left[(\text{tr}Q)^2 + \epsilon\,\text{tr}Q^2\right]; \qquad (10.21)$$

where

$$\epsilon = +1 \text{ for Sp (n)}$$
$$\epsilon = -1 \text{ for SO (n)}.$$

A few comments about (10.21) are in order. Note that Q is equivalent to $\exp N$, and so we could expand both sides of (10.21) in power series in N in order to extract the desired coefficient e_i. The procedure yields

$$\begin{aligned}
\text{Tr}\,N^2 &= (A + 2\epsilon)\,\text{tr}\,N^2, \\
\text{Tr}\,N^4 &= (A + 8\epsilon)\,\text{tr}\,N^4 + 3\,(\text{tr}\,N^2)^2, \\
\text{Tr}\,N^6 &= (A + 32\epsilon)\,\text{tr}\,N^6 + 15\,\text{tr}\,N^2\,\text{tr}\,N^4.
\end{aligned} \qquad (10.22)$$

What is the analog of formula (10.22) for E_8? The fundamental irreducible representation of E_8 decomposes under its subgroup SO (16):

$$\underline{248} = 120 + 128_{(+)}.$$

120 is the adjoint of SO (16), and $128_{(+)}$ is none other than the positive chirality spinor of SO (16). E_8 and SO (16) both have rank eight, so the weights of their representations are eight-dimensional vectors. Let $e_i \cdot e_j = \delta_{ij}$ denote the standard basis; the weight of the fundamental irreducible representation of SO (16) are $\pm e_i$, $i = 1, \cdots, 8$. Consequently, the weights of the adjoint representation read:

$$\pm e_i \pm e_j, \quad i < j, \tag{10.23}$$

to which should be added eight zeros corresponding to the Cartan subalgebra. For the spinor, we have:

$$\begin{aligned} &\tfrac{1}{2} \sum_{a=1}^{8} \epsilon_a e_a, \\ &\epsilon_a = \pm 1 \\ &\prod_{a=1}^{8} \epsilon_a = +1. \end{aligned} \tag{10.24}$$

A given matrix N in the adjoint representation of E_8 can be written in terms of eight parameters, $x_1, \cdots, x_8$. Thus, the character of the 248 representation of E_8 simply corresponds to the sum of 120 and 128 of SO (16):

$$\begin{aligned} \mathrm{Tr}_{E_8} Q &= \sum_{i<j} \exp\left(\pm x_i \pm x_j\right) + \sum_\epsilon \exp\left(\tfrac{1}{2} \sum_{a=1}^{8} \epsilon_a x_a\right) \\ &= \tfrac{1}{2} \left[\left(\sum_i 2\cosh x_i\right)^2 - \left(\sum_i 2\cosh 2x_i\right)\right] \\ &\quad + \tfrac{1}{2} \left[\prod_{i=1}^{8} 2\cosh \tfrac{x_i}{2} + \prod_{i=1}^{8} 2\sinh \tfrac{x_i}{2} \right]. \end{aligned} \tag{10.25}$$

The following expansion applies to E_8:

$$\begin{aligned} \mathrm{Tr}\, N^4 &= \tfrac{1}{100} \left(\mathrm{Tr}\, N^2\right)^2, \\ \mathrm{Tr}\, N^6 &= \tfrac{1}{7200} \left(\mathrm{Tr}\, N^2\right)^3. \end{aligned} \tag{10.26}$$

The relations (10.22) and (10.26) are crucial ingredients in the Green-Schwarz anomaly cancellation, to which we now dedicate the reminder of this exposé. The only groups for which the anomalies cancel are SO (32) and $E_8 \times E_8$ [5]. As pointed out earlier, these groups are the internal symmetries required to describe the superstring in ten dimensions. Anomalies of this scale manifest themselves primarily in hexagon diagrams. A general rule for

$2n$ dimensions is that the first dangerous diagram is an ($n - 1$-gon). This is easily understood from the fact that the anomalous divergence of a Yang-Mills gauge current consists of

$$\partial \cdot J^a \sim \epsilon^{\mu_1 \cdots \mu_{2n}} \, \text{tr} \left(\Lambda^a \, F_{\mu_1 \mu_2} \cdots F_{\mu_{2n-1} \mu_{2n}} \right). \qquad (10.27)$$

This algebra in turn is a function of the chiral fermions which go through the loop. Analogous gravitational anomalies can occur, particularly for odd n, in the energy-momentum tensor. The standard $N = 2$ supergravity theories in ten dimensions are known to be anomaly-free [4]. In one of the theory's incarnations, namely the $2A$, this is due to the trivial cancellations between contributions of left-handed and right-handed fermions, whereas in the case of $2B$, it is a consequence of highly non trivial cancellations discovered by the authors in reference [4]. $N = 1$ theories present even more difficult challenges. The anomaly cancellation in type I SO(32) superstring theory was discovered by evaluating hexagon loop diagrams in string theory [20].

Consider the effective action $\mathcal{L}_{\text{eff}}$ of the low-energy expansion of string theory. $\mathcal{L}_{\text{eff}}$ is defined over fields of massless modes, which we write as Φ_0; it is obtained by integrating out all the fields associated with massive string modes, the Φ_m:

$$e^{i\mathcal{L}_{\text{eff}}(\Phi_0)} \sim D\Phi_m \, e^{i\mathcal{L}_{\text{string}}(\Phi_0, \Phi_m)}. \qquad (10.28)$$

$\mathcal{L}_{\text{eff}}$ can be expanded in a series of operators of increasing dimension. The leading terms correspond to the point-particle theory for $N = 1$, $D = 10$ super Yang-Mills theory coupled with supergravity. The higher order terms represent string corrections whose effects are indeed suppressed at low energy E_{low} by powers of E_{low}/m, where m is the characteristic mass scale of the string. Green and Schwarz have found that the theory is known to be anomalous for every Yang-mills group, unless terms not present in the minimal point-particle theory are taken into account [5].

Massless fields consist of a super Yang-Mills multiplet and a supergravity multiplet. The super Yang-Mills multiplet contains vector fields $A^a{}_\mu$ and left-handed Majorana-Weyl spinor fields χ_L^a. The index a takes n values corresponding to the generators of the Yang-Mills group G; incidentally, $n = \dim G$. The supergravity multiplet contains a graviton $g_{\mu\nu}$, a second-rank antisymmetric tensor field $B_{\mu\nu}$, a scalar ρ, a left-handed Majorana-Weyl gravitino $\Psi_{\mu L}$, and a right-handed Majorana-Weyl spinor λ_R. These fields

are all singlets of the gauge group. The chiral spinors χ, ρ, and λ going
through the hexagon (and higher) diagrams give rise to gauge, gravitational,
and mixed anomalies.

The use of differential forms is an efficient means to investigate the anomalies. To gauge potential, we associate 1-forms:

$$
\begin{aligned}
A &\equiv A^a{}_\mu\, \lambda^a\, dx^\mu \\
\omega &= \omega_\mu\, dx^\mu.
\end{aligned}
\tag{10.29}
$$

ω_μ, a spin connection, is a 10×10 matrix in the fundamental representation
of the Lorentz algebra $SO\,(9,1)$; the λ^a are $n \times n$ matrices in the adjoint
representation of the Yang-Mills algebra, that is, they form the algebra's
structure constant. The two-form fields strength reads

$$
\begin{aligned}
F &\equiv \tfrac{1}{2} F_{\mu\nu}\, dx^\mu \wedge dx^\nu = dA + A^2 \\
R &\equiv \tfrac{1}{2} R_{\mu\nu}\, dx^\mu \wedge dx^\nu = d\omega + \omega^2.
\end{aligned}
\tag{10.30}
$$

A few notations are in order. F stands for the Yang-Mills field strength while
R denotes the Riemannian curvature tensor. Infinitesimal gauge transformations with parameters α and β satisfy the relations:

$$
\begin{aligned}
\delta A &= d\alpha + [A, \alpha] \\
\delta\omega &= d\beta + [\omega, \beta],
\end{aligned}
\tag{10.31}
$$

and

$$
\begin{aligned}
\delta F &= [F, \alpha] \\
\delta R &= [R, \beta].
\end{aligned}
\tag{10.32}
$$

It is now useful to introduce the Chern-Simons 3-forms

$$
\begin{aligned}
\omega_{3Y} &= \mathrm{Tr}\,(AF - \tfrac{1}{3}A^3) \\
\omega_{3L} &= \mathrm{tr}\,(\omega R - \tfrac{1}{3}\omega^3),
\end{aligned}
\tag{10.33}
$$

which have the properties

$$
\begin{aligned}
d\omega_{3Y} &= \mathrm{Tr}\,F^2 \\
d\omega_{3L} &= \mathrm{tr}\,R^2,
\end{aligned}
\tag{10.34}
$$

and

$$
\begin{aligned}
\delta\omega_{3Y} &= -d\omega'_{2Y} \\
\delta\omega_{3L} &= -d\omega^1_{2L}.
\end{aligned}
\tag{10.35}
$$

According to the notations of the authors in [5], the ω_{2L}^1 and ω_{2L}^1 stand for the two-forms while the superscript 1 indicates that they are linear in the infinitesimal parameters α and β respectively.

Anomalies arising from loops of chiral fermions are easily detected by 12-forms. From this, one extracts the actual 10-forms whose integral is the anomaly. This is done by replacing $F \rightarrow F + \alpha$ and $R \rightarrow R + \beta$ and by extracting in a second step the terms linear in α or β. The final result is

$$
\begin{aligned}
I_{12} \propto{} & -\tfrac{1}{15} \operatorname{Tr} F^6 - \tfrac{1}{960} \operatorname{Tr} F^2 \left(4\operatorname{tr} R^4 + 5(\operatorname{tr} R^2)^2\right) \\
& + \tfrac{1}{24} \operatorname{Tr} F^4 \operatorname{tr} R^2 + \left(\tfrac{1}{32} + \tfrac{n-496}{13,824}\right) (\operatorname{tr} R^2)^3 \\
& + \left(\tfrac{1}{8} + \tfrac{n-496}{5760}\right) \operatorname{tr} R^2 \operatorname{tr} R^4 + \left(\tfrac{n-496}{7560}\right) \operatorname{tr} R^6 .
\end{aligned} \tag{10.36}
$$

The big numbers originate from characteristic classes. We will say more about these in the following chapter. For a detailed account of 12-forms characterizing all the gauge, gravitational and mixed anomalies due to χ, Ψ, and λ loops, the reader is referred to [4]. Meanwhile, Green and Schwarz ask the following question: is it possible for $\mathcal{L}_{\text{eff}}$ to contain a non-gauge-invariant local interaction term $\mathcal{L}_c$ whose gauge variation $\delta\mathcal{L}$ cancels the anomaly associated with I_{12}? They discovered in [5] that this is only possible if I_{12} factorizes into an expression of the form:

$$
I_{12} = \left(\operatorname{tr} R^2 + k \operatorname{tr} F^2\right) X_8. \tag{10.37}
$$

X_8 is an 8-form constructed in terms of F and R. The necessary and sufficient condition for this to hold is

$$
n = \dim G = 496. \tag{10.38}
$$

When satisfied, the condition gives the following value for k:

$$
k = -\frac{1}{30}, \tag{10.39}
$$

and

$$
\begin{aligned}
X_8 ={} & \tfrac{1}{24} \operatorname{Tr} F^4 - \tfrac{1}{7200} (\operatorname{Tr} F^2)^2 \\
& - \tfrac{1}{240} \operatorname{Tr} F^2 \operatorname{tr} R^2 \\
& + \tfrac{1}{8} \operatorname{tr} R^4 + \tfrac{1}{32} (\operatorname{tr} R^2)^2 .
\end{aligned} \tag{10.40}
$$

Factorizing I_{12} in (10.37) tell us that to cancel the anomaly, we need only add the form

$$
\mathcal{L} \sim \int B \wedge X_8, \tag{10.41}
$$

B is a 2-form congruent to $B_{\mu\nu}\,dx^\nu \wedge dx^\nu$ with gauge transformation

$$\delta B = \omega_{2L}^1 - \frac{1}{30}\,\omega_{2Y}^1.$$ (10.42)

The formulae (10.38) and (10.39) have only two known solutions: $SO\,(32)$ and $E_8 \times E_8$. An exhaustive analysis of the artithmetic of the present finding has been carried out in reference [5] and [20]. We will not repeat them in this chapter, referring interested readers to these two references. Note that additional solutions exit, namely for $[U\,(1)]^{496}$ and $E_8 \times [U\,(1)]^{248}$. Unfortunately they do not correspond to consistent string theories and this makes them less attractive. $E_8 \times E_8$ describe, to date, the most promising string theory, the so-called heterotic string theory discovered a decade ago by David Gross and collaborators [6].

10.4   References

[1] Schwinger, J.: Physical Reviews 82 (1951) 664.

– Adler, S. L.; Lee, B. W.; Treiman, S. B.; and Zee, A.: Physical Reviews D4 (1971) 3497.

– Callan, C.; Dashen, R. and Gross, D. J.: Phys. Letters B63 (1976) 334.

[2] Langacker, P.: Physical Rep. 72 (1981) 185.

– Zee, A.: in **Unity of Forces in the Universe**, Vol. I and II, World Scientific, 1982.

[3] Salam, A. and Strathdee, J.: Annals of Physics 141 (1982) 316.

– Nieuwenhuizen, P.:*An Introduction to Simple Supergravity and the Kaluza-Klein Program*, in **Relativity Groups and Topology**, II, Ed. R. Stora, North Holland 1985.

[4] Alvarez-Gaumé, L and Witten, E.: *Gravitational Anomalies*, Nuclear Physics B234 (1983) 269.

[5] Green, M. B. and Schwarz, J. H.: Physics Letters B149 (1984) 117.

[6] Gross, D. J.: *The Heterotic String*, in **Unified String Theories**, Eds. Gross, D. J. and Green, M. B. 357-399, World Scientific 1986.

[7] Atiyah, M. F. and Singer, I. M.: Annals of Mathematics 87 (1968), 485-546. Idem, 93 (1971) 119-139.

[8] Jackiw, R.; Nohl, C. and Rebbi, L.: in **Particles and Fields**, Eds. D. Boch and A. Kamal, Plenum Press 1978.

– Nielsen, N. K.; Römer, H. and Schoer, B.: Nuclear Physics B 136 (1978) 478.

[9] Atiyah, M. F.; Patodi, V. I. and Singer, I. M.: Mathematical Proceedings Cambridge Philosophical Society 77 (1975) 43.

–(Idem), 78 (1975), 405 and 79 (1976) 71.

[10] Atiyah, M. F. and Singer, I. M.: in Proc. National Academy of Sciences USA 81 (1984) 2597.

[11] Alvarez-Gaumé, L. and Ginsparg, P.: Nuclear physics B243 (1984) 449.

– (Idem) Annals of Physics 161 91985) 423.

[12] Alvarez, O., Singer, I. M. and Zumino, B: Communications Mathematical Physics 96 (1984) 409.

[13] Witten, E.: Nuclear Physics B223 (1983) 422.

[14] Faddeev, L. D.: Physics Letters B 145 (1984) 81.

– Faddeev, L. D. and Shatashvili, S.: Math. Phys. 60 (1984) 206.

[15] Zumino, B.: Nuclear Physics B253 91985) 477.

[16] Nelson, P. and Alvarez-Gaumé, L.: Communications Mathematical Physics 99 (1985) 103.

[17] Witten, E.: *An* SU (2) *Anomaly*, Phys. Letters B117 (1982) 324.

[18] Witten, E.: *Global Gravitational Anomalies*, Commun. Mathematical Physics 100 (1985) 197.

[19] Weyl, H.: **Classical Groups**, Princeton University Press 1939.

[20] Green, M. B. and Schwarz, J. H.: Nuclear Physics B243 (1984) 285.

– Green, M.; Schwarz, J. H., and Witten, E.: **Superstring Theory**, Cambridge University Press, Vol. I and II, New York, 1987.

[21] Wess, J. and Bagger, J.: **Supersymmetry and Supergravity**,

Princeton Series in Physics, Princeton University Press, Princeton, NJ 1983.

Chapter 11

Anomalies and the Index Theorem

The most suitable topological framework for the study of anomalies is encompassed in the theory of fiber bundles to which we devoted Chapters 6 and 7: anomalies are known to occur whenever the determinant bundle of a suitable operator is nontrivial over the orbit space [1, 2, 3]. Much of this depends on the relative twisting of the line bundle. In turn, such twistings are computed via the family of index theorem, and reflect, on the physical side, the fact that the conservation of anomalous fermionic currents does not survive quantization [4]. The index theorem is essentially a bridge between the analytical properties of differential operators on fiber bundles and the topological properties of the fiber bundles themselves. The most readily available example of such is the Gauss-Bonnet theorem, which relates the number of harmonic forms on a given manifold, M, (i.e. the so-called Betti numbers) to the Euler character given by integrating the Euler form over M. The relevant differential operator here is referred to as the exterior derivative mapping,

$$\mathbb{C}^\infty\left(\Lambda^p\right) \;\to\; \mathbb{C}^\infty\left(\Lambda^{p+1}\right),$$

and the analytic property to which we alluded is the number of zero-frequency solutions to a generalized Laplace equation.

The index theorem is useful in exhibiting general analogs of the Gauss-Bonnet theorem for other differential operators. The index of a given op-

163

erator is often determined by the number of zero-frequency solutions to a generalized Laplace equation and is expressed in terms of the characteristic classes of the fiber bundles involved. It is, moreover, a powerful tool in the analysis of global gauge and gravitational anomalies. The present chapter is, therefore, a prerequisite for any study of global anomalies in general. We will provide here the general formulation of the index theorem and discuss its relationship with chiral and gravitational anomalies. The use of the index theorem for global gauge and gravitational anomalies will be the topic of upcoming chapters. We shall follow the presentation outlined in reference [5] by Alvarez-Gaumé, and in reference [6] by Eguchi and collaborators.

11.1   The Index Theorem

The starting point is a Dirac operator for fermions coupled to external gauge and gravitational fields in even dimensions. To this operator is associated two vector bundles over M, which we write as $R_+ \otimes V$, and $R_- \otimes V$. The space V carries the representation of the gauge group, while R_+ (R_-) denotes the space of positive (respectively negative) chirality spinors. A Weyl operator,

$$D_+ = i \, \displaystyle{\not}D \, p_+,$$

(whose adjoint is $D_- = \displaystyle{\not}D \, p_-$) sends objects localized in $R_+ \otimes V$ into objects in $R_- \otimes V$, thanks to the formula

$$R_+ \otimes V \underset{D_-}{\overset{D_+}{\rightleftarrows}} R_- \otimes V; \tag{11.1}$$

p is a closed form. The explicit form for $D_\pm$ is

$$D_\pm = i\gamma^\mu \left(\partial_\mu \frac{1}{2}\omega_{\mu ab} \Sigma^{ab} + A_\mu\right) P_\pm. \tag{11.2}$$

The index of D_+ is the dimension of the kernel of D_+ minus the kernel of $D_+^\dagger = D_-$:

$$\mathrm{ind}\, D_+ = \dim \ker D_+ - \dim \ker D_-. \tag{11.3}$$

Atiyah and Singer [7] showed that $\mathrm{ind}\, D_+$ is a number which depends only on the topological set up (11.1) and it is given by the integral over M

of a particular characteristic class. For instance, for the operator (11.2) the index is

$$\text{ind}\, D_+ \;=\; \int_M \left[\text{ch}\,(F)\,\hat{A}\,(M)\right]_{\text{vol}}, \tag{11.4}$$

where

$$\hat{A}\,(M) \equiv \prod_a \frac{x_a/2}{\sinh x_a/2},$$

and

$$\text{ch}\,(F) \;=\; \text{Tr}\, e^{iF/2\pi}.$$

$\hat{A}\,(M)$, the $\hat{A}$ Dirac genus of M, is a polynomial in the 2-form x_a; it is a finite polynomial since M is finite dimensional. The subscript *vol* in formula (11.4) means that one has to extract the form whose degree is equal to the dimension of M. There is an expansion of $\hat{A}\,(M)$ that makes use of polynomials in the Pontrjagin classes, namely:

$$
\begin{aligned}
\hat{A}\,(M) \;&=\; 1 + \tfrac{1}{(4\pi)^2}\tfrac{1}{12}\,\text{Tr}\, R^2 + \tfrac{1}{(4\pi)^6}\left[\tfrac{1}{288}\,(\text{Tr}\, R^2)^2 + \tfrac{1}{360}\text{Tr}R^4\right] \\
&\quad + \tfrac{1}{(4\pi)^6}\left[\tfrac{1}{10368}\,(\text{Tr}\, R^2)^3 + \tfrac{1}{4320}\,\text{Tr}\, R^2\,\text{Tr}\, R^4 + \tfrac{1}{5670}\,\text{Tr}\, R^6\right] \\
&\quad + \tfrac{1}{(4\pi)^8}\left[\tfrac{1}{497664}\,(\text{Tr}\, R^2)^4 + \tfrac{1}{103680}\,(\text{Tr}R^2)^2 + \text{Tr}\, R^4\right. \\
&\quad \left. + \tfrac{1}{68040}\,\text{Tr}\, R^2\,\text{Tr}\, R^6 + \tfrac{1}{259200}\,(\text{Tr}\, R^4)^2 + \tfrac{1}{75600}\,\text{Tr}\, R^8\right] + \cdots \\
&=\; 1 + \tfrac{1}{2^2}\left(-\tfrac{1}{6}p_1\right) + \tfrac{1}{2^4}\left(\tfrac{7}{360}p_1^2 - \tfrac{1}{90}p_2\right) \\
&\quad + \tfrac{1}{2^6}\left(-\tfrac{31}{15120}p_1^3 + \tfrac{11}{3780}p_1 p_2 - \tfrac{1}{945}p_3\right) \\
&\quad + \tfrac{1}{2^8}\left(\tfrac{127}{604800}p_1^4 - \tfrac{113}{226800}p_1^2 p_2 + \tfrac{4}{14175}p_1 p_3 + \tfrac{13}{113400}p_2^2 - \tfrac{1}{9450}p_4\right) + \cdots
\end{aligned}
\tag{11.5}
$$

When $\dim R = n$, we have

$$\text{ch}(F) \;=\; n + \frac{i}{2\pi}\,\text{Tr}\, F + \frac{i^2}{2(2\pi)^2}\,\text{Tr}F^2 + \cdots + \frac{i^n}{n!(2\pi)^n}\,\text{Tr}\, F_+^n + \cdots \tag{11.6}$$

One can then compute the index theorem by combining equations (11.5) and (11.6). In dimensions four and eight the index is

$$
\begin{aligned}
\text{ind}\, D_+ \;&=\; \tfrac{1}{(2\pi)^2}\int_M \left(\tfrac{i^2}{2}\,\text{Tr}\, F^2 + \tfrac{n}{48}\,\text{Tr}\, R^2\right) \qquad d = 4 \\
\text{ind}\, D_+ \;&=\; \tfrac{1}{(2\pi)^4}\int_M \left(\tfrac{i^4}{24}\,\text{Tr}\, F^4\right) + \tfrac{i^2}{96}\,\text{Tr}F^2\,\text{Tr}\, R^2 \\
&\quad + \tfrac{n}{4608}\,(\text{Tr}R^2)^2 + \left(\tfrac{n}{5760}\,\text{Tr}\, R^4\right) \qquad d = 8.
\end{aligned}
\tag{11.7}
$$

Let us focus now on the operators whose indices are essential in the computation of anomalies. These are obtained by replacing R by some particular vector bundle. In order to illustrate this point, we consider the case of the graviton field $R = TM$, where TM is the tangent bundle over M. In this instance, A is simply the spin connection taking values on the Lie algebra of $SO(2n)$ in the vector representation,

$$(T^{ab})_{cd} = \delta^a{}_c \delta^b{}_d - \delta^a{}_d \delta^b{}_c.$$

Thus, we get

$$\operatorname{Tr} e^{(R_{ab} T^{ab}/4\pi)} = \sum_a 2 \cosh x_a. \tag{11.8}$$

The quantization of a spin $\frac{3}{2}$ field requires adding ghost fields if one is to remove unphysical degrees of freedom. When including the ghost field for a spin $\frac{3}{2}$ field, the index theorem is

$$\operatorname{ind} i \, \slashed{D}_{3/2} = \int \hat{A}(M) \left(\operatorname{Tr} e^{R/2\pi} - 1 \right) \operatorname{ch}(F). \tag{11.9}$$

The last factor in (11.9) accounts for the possibility that the spin 3/2 field carries extra gauge indices. There is a dimensional dependence of $\operatorname{Tr} e^{R/2\pi}$, as shown by the term $\hat{A}(M) \left(\operatorname{Tr} e^{R/2\pi} - 1 \right)$. To order 16, the polynomial in question is

$$
\begin{aligned}
\hat{A}(M) \left(\operatorname{Tr} e^{R/2\pi} - 1 \right) =\ & -\tfrac{1}{(4\pi)^2} 2\operatorname{Tr} R^2 + \tfrac{1}{(4\pi)^4}\left[-\tfrac{1}{6}\left(\operatorname{Tr} R^2\right)^2 + \tfrac{2}{3}\operatorname{Tr} R^4 \right] \\
& + \tfrac{1}{(4\pi)^6}\left[-\tfrac{1}{144}\left(\operatorname{Tr} R^2\right)^3 + \tfrac{1}{20}\operatorname{Tr} R^2 \operatorname{Tr} R^4 - \tfrac{4}{45}\operatorname{Tr} R^6 \right] \\
& + \tfrac{1}{(4\pi)^8}\left[-\tfrac{1}{5184}\left(\operatorname{Tr} R^2\right)^4 + \tfrac{1}{540}\left(\operatorname{Tr} R^2\right)^2 \operatorname{Tr} R^4 \right. \\
& \left. \qquad -\tfrac{22}{2835}\operatorname{Tr} R^2 \operatorname{Tr} R^6 + \tfrac{1}{540}\left(\operatorname{Tr} R^4\right)^2 + \tfrac{2}{315}\operatorname{Tr} R^8 \right] + \cdots \\
=\ & \tfrac{1}{2^2}(4p_1) + \tfrac{1}{2^4}\left(\tfrac{2}{3}p_1^2 - \tfrac{8}{3}p_2\right) \\
& + \tfrac{1}{2^6}\left(\tfrac{1}{30}p_1^3 - \tfrac{2}{15}p_1 p_2 + \tfrac{8}{15}p_3\right) \\
& + \tfrac{1}{2^8}\left(\tfrac{1}{1260}p_1^4 - \tfrac{16}{945}p_1^2 p_2 - \tfrac{8}{189}p_1 p_3 + \tfrac{52}{945}p_2^2 - \tfrac{16}{315}p_4\right) \\
& + \cdots
\end{aligned}
$$
$$\tag{11.10}$$

In even dimensions, a pair of chiral spinor representations whose self-dual representation is $SO(2n)$ appears along with a number of anomaly-free representations [8]. In light of this, it is important to consider the index for a bi-spinor field $\Phi_{\alpha\beta}$. The index theorem for such a field is expressed in terms of the signature theorem [9]. Signatures are computed with respect to Betti numbers. The starting point is to write down harmonic forms in the de Rham complex; they decompose into self-dual and anti-self dual pieces. There is another approach which consists of defining the signature $\sigma(M)$ in terms of the heat kernel expansion:

$$\sigma(M) = \text{Tr} \star e^{-\beta \mathcal{L}}; \tag{11.11}$$

in this formula, $\star$ is the Hodge duality operator and $\mathcal{L} : \Lambda_p \rightarrow \Lambda_p$. (The zeroes of $\mathcal{L}$ are called harmonic forms.) Index theory makes use of the fact that the space of harmonic forms is equivalent to that of the closed and co-closed forms. Formula (11.11) is related to the Atiyah-Singer theorem as follows: it can be computed as an index problem for a Dirac-like operator interpolating between

$$R_+ \otimes (R_+ + R_-) \rightarrow R_- \otimes (R_+ \otimes R_-). \tag{11.12}$$

Combining (11.7) with the additivity of the Chern character yields

$$\begin{aligned} \sigma(M) &= \int_M \text{ch}(R_+ \oplus R_-) \, \widehat{A}(M) \\ &= \int_M (\text{ch} R_+ + \text{ch} R_-) \, \widehat{A}(M). \end{aligned} \tag{11.13}$$

The next term to be computed is the character of SO $(2n)$ in the representation $R_+ \oplus R_-$:

$$\text{Tr} \exp \frac{1}{2\pi} R_{ab} \Sigma^{ab} = \text{ch} R_+ + \text{ch} R_-.$$

The computation reveals that

$$\text{ch} R_+ + \text{ch} R_- = \prod_{a=1}^{n} 2\cosh \frac{x_a}{2},$$

this allows one to obtain the sought signature, namely,

$$\sigma(M) = \int_M [\mathcal{L}(M)]_{\text{vol}}; \tag{11.14}$$

where

$$\mathcal{L}(M) = \prod_a 2 \frac{x_a/2}{\tanh x_a/2}.$$

$\mathcal{L}(M)$ denotes the Hirzebruch $\mathcal{L}$-polynomial of M, whose expansion is

$$
\begin{aligned}
\mathcal{L}(M) &= 1 - \frac{1}{(2\pi)^2} \frac{1}{6} \operatorname{Tr} R^2 + \frac{1}{(2\pi)^4} \left[\frac{1}{72} \left(\operatorname{Tr} R^2 \right)^2 - \frac{7}{180} \operatorname{Tr} R^4 \right] \\
&\quad + \frac{1}{(2\pi)^6} \left[-\frac{1}{1296} \left(\operatorname{Tr} R^2 \right)^3 + \frac{7}{1080} \operatorname{Tr} R^2 \operatorname{Tr} R^4 - \frac{31}{2835} \operatorname{Tr} R^6 \right] \\
&\quad + \frac{1}{(2\pi)^8} \left[\frac{1}{31104} \left(\operatorname{Tr}^2 \right)^4 - \frac{7}{12960} \left(\operatorname{Tr} R^2 \right)^3 \operatorname{Tr} R^4 \right. \\
&\quad \left. + \frac{31}{17010} \operatorname{Tr} R^2 \operatorname{Tr} R^6 + \frac{49}{64800} \left(\operatorname{Tr}^4 \right)^2 - \frac{127}{37800} \operatorname{Tr} R^8 \right] + \cdots \\
&= 1 + \frac{1}{3} p_1 + \left(-\frac{1}{45} p_1^2 + \frac{7}{45} p_2 \right) \\
&\quad + \left(\frac{2}{945} p_1^3 - \frac{13}{945} p_1 p_2 + \frac{62}{945} p_3 \right) \\
&\quad + \left(-\frac{1}{4725} p_1^4 + \frac{22}{14175} p_1^2 p_2 - \frac{71}{14175} p_1 p_2 - \frac{19}{14175} p_2^2 + \frac{127}{4725} p_4 \right) + \cdots
\end{aligned}
$$

$$(11.15)$$

The Atiyah-Patodi-Singer theorem (APS) deals with compact manifolds with boundaries. The index theorem (11.7), in this case, generalizes to

$$
\operatorname{ind} D\!\!\!/ = \int_M \widehat{A}(M) \operatorname{ch}(F) - \frac{1}{2} \left(\eta(0) + h \right). \tag{11.16}
$$

The first term in (11.6) is the standard volume term, and the second term the boundary correction; $\eta(0)$ is a spectral invariant for the on the boundary of M (we refer the interested reader to Chapter 1 for a thorough definition of η). It is finite and measures the spectral asymmetry of $D\!\!\!/^{2n-1}$. The term h, in (11.16), counts the number of zero modes of $D^{(2n-1)}$ on ∂M. Applications of the APS index theorem in the study of global anomalies appeared in [10] and [11] and will be discussed further in upcoming chapters.

11.2   Characteristic Classes

11.2.1   The Chern Character

As we have seen earlier, index theory requires the use of characteristic classes. These are actually essential tools in the formulation of the index theorem. The Whitney sum of bundles and the tensor product of bundles are among the most frequently used such tools. The total Chern class works well for the Whitney sum; it is however, unreliable for product bundles. The total Chern class:

$$
c(\mathcal{L}) = \prod_i (1 + x_j), \quad \mathcal{L} = \text{vector bundle},
$$

needs to be reformulated in terms of polynomials in the $\{x_i\}$. One such polynomial is the Chern character, $\mathrm{ch}\,(\mathcal{L})$. It is defined by the invariant polynomial:

$$\mathrm{ch}\,(\alpha) \;=\; \mathrm{Tr}\,exp\,(\frac{i}{2\pi}\alpha) \;=\; \sum_i \frac{1}{i}\,\mathrm{Tr}\,(\frac{i}{2\pi}\alpha)^i. \qquad (11.17)$$

The Chern character of the line bundle $\mathcal{L}$ can be expanded in terms of Chern classes:

$$\mathrm{ch}\,(\mathcal{L}) \;=\; \sum_{i=1}^{k} e^{x_i} \;=\; \underbrace{k}_{=\dim\,\mathrm{of}\,\mathcal{L}} + c_1\,(E) + \frac{1}{2}\,(c_1^2 - 2c_2)\,(E) + \cdots \qquad (11.18)$$

11.2.2   Hirzebruch Polynomial and Pontrjagin Classes

The Hirzebruch $\mathcal{L}$-polynomial

$$L\,(\mathcal{L}) \;=\; \prod_i \frac{x_i}{\tanh x_j}, \qquad (11.19)$$

is a multiplicative characteristic class, characteristic of the signature index formula. The $\hat{A}$ polynomial, whose use centers in the computation of the spin index formula, is given by the formula

$$\hat{A}\,(\mathcal{L}) \;=\; \prod_i \frac{x_i/2}{\sinh\,(x_i/2)}. \qquad (11.20)$$

The total Pontrjagin class (of a real $O\,(n)$ bundle) $\mathcal{L}$ with curvature Ω is given by the invariant polynomial

$$p\,(\mathcal{L}) \;=\; \det\,(I - \frac{1}{2\pi}\,\Omega) \;=\; 1 + p_1 + p_2 + \cdots \qquad (11.21)$$

I is the identity matrix.

When dealing with index theory, it is often convenient to express the Pontrjagin classes of a real bundle in terms of the Chern classes of complex bundles. This procedure employs the relation

$$\mathcal{L}_{\mathrm{complex}} \;=\; \mathcal{L} \otimes \mathbb{C},$$

that is, the complexification of $\mathcal{L}$. Complexification arises naturally as an inclusion of $GL(n, \mathbb{R})$ into $GL(n, \mathbb{C})$. Consider a skew-adjoint real matrix, Y. The relation

$$\det\left(I + \frac{i}{2\pi}Y\right) = 1 - p_1(Y) + p_2(Y)\cdots$$

yields

$$p_n(\mathcal{L}) = (-1)^n c_{2n}(\mathcal{L}_{\text{complex}}). \tag{11.22}$$

We can repeat the converse by forgetting the complex structure on $\mathcal{L}$ using the so-called forgetful functor. Hence,

$$(\mathcal{L}_{\text{real}})_{\text{complex}} = \mathcal{L} \oplus \tilde{\mathcal{L}}.$$

$\tilde{\mathcal{L}}$ stands for the dual (isomorphic) bundle to $\mathcal{L}$. Using the above formula, we have

$$c(\tilde{\mathcal{L}}) = 1 - c_1(\mathcal{L}) + c_2(\mathcal{L}) - c_3(\mathcal{L})\cdots$$

which gives rise to

$$\begin{aligned}
c(\mathcal{L}_{\text{complex}}) &= 1 - p_1(\mathcal{L}_{\text{real}}) + p_2(\mathcal{L}_{\text{real}}) - \cdots \\
&= c(\mathcal{L})\,c(\tilde{\mathcal{L}}) \\
&= [1 + c_1(\mathcal{L}) + c_2(\mathcal{L}) + \cdots] \\
&\quad \times [1 - c_1(\mathcal{L}) + c_2(\mathcal{L}) + \cdots].
\end{aligned} \tag{11.23}$$

Accordingly,

$$\begin{aligned}
p_1(\mathcal{L}_{\text{real}}) &= (c_1^2 - 2c_2)(\mathcal{L}) \\
p_2(\mathcal{L}_{\text{real}}) &= (c_2^2 2c - 1c_3 + 2c_4)(\mathcal{L}).
\end{aligned} \tag{11.24}$$

Anomalies essentially measure the non-triviality of the determinant line bundle of a family of Dirac operators, which is given topologically by the first Chern class. As discussed above, in the case of real bundles, one can derive a local formula for the Chern class of $\mathcal{L}$ in terms of characteristic classes. Global anomalies, on the other hand, are detected by $c_1(\mathcal{L})$ and this generalizes the whole issue to the complex case. Anomalies do have another geometrical interpretation, namely, the determinant bundle carries a connection, and the anomaly is represents nothing other than its curvature and holonomy [12, 13, 14].

11.3   The Index Theorem and Anomalies

Consider a compact $2k$-dimensional manifold, M^{2k}, to which is associated an infinite dimensional affine space Ξ, the space of all possible gauge fields. Let Ξ/Γ denote the space of gauge-invariant configurations. The term Γ stands for the gauge transformations which obey Gauss' law. The existence of a fermion determinant defines a complex line bundle over Ξ. The case in which this line bundle is non-trivial is important for a study of the anomalies. In previous chapters, namely Chapters 6 and 7, we have seen that line bundles are classified by their first Chern class (the monopole number in physics terms), which belongs to $H^2(M^{2k};\mathbb{Z})$. Hence, the 2-form

$$c_1 = \frac{1}{(2\pi)^{2k+1}(k+1)!} \int_{M^{2k}} \operatorname{Tr} F^{k+1}, \qquad (11.25)$$

is a function of $H^2(\Xi/\Gamma;\mathbb{R})$. The generators of $H^2(\Xi/\Gamma)$ are 2-cycles deduced from 2-parameter families of gauge fields; the 2-cycle in Ξ/γ on which we integrate c_1 is a 2-sphere.

The use of the index theorem for families of elliptic operators is therefore warranted. Roughly speaking, we choose a theory living in M^{2k}, with gauge group G. In a given region of M^{2k}, we write $\nabla \in \Xi$ for all possible connections. Because of the compactness of M^{2k}, there is a finite number of such possible connections. To each ∇ we associate the Weyl operators $D_+(\nabla)$, $D_-(\nabla)$. As a consequence, we have an infinite dimensional parameter family of Dirac operators which are parametrized by ∇. Such a family can be parametrized by Ξ/Γ owing to the relation $D_\pm(g^{-1}\nabla g) = g^{-1} D_\pm(\nabla) g$. The procedure relies on the elliptic self-adjoint operators $D_-(\nabla) D_+(\nabla)$ and $D_+(\nabla) D_-(\nabla)$ whose eigenfunctions span the Hilbert spaces $\mathcal{H}_+(\nabla)$ and $\mathcal{H}_-(\nabla)$ respectively, and a Hilbert bundle over Ξ/Γ. The term $\operatorname{ind} D_\nabla$ is a constant independent of ∇. But it is also the simplest topological invariant that one can define out of this infinite dimensional family of operators.

Next, we consider a finite dimensional subset K of Ξ which projects to some compact set in Ξ/Γ. On K, we study of the two Hilbert bundles $\mathcal{H}_K^+$ and $\mathcal{H}_K^-$. The eigenvalues of $L_-(p) = D_+(\nabla_p) D_-(\nabla_p)$ change with respect to p whenever one moves ∇_p in the subset K. The eigenvalues

$$\lambda_0(p) \leq \lambda_1(p) \leq \cdots \leq \lambda_n(p) \leq \cdots$$

split $\mathcal{H}^+\left(\nabla_p\right)$ and $\mathcal{H}^-\left(\nabla_p\right)$ into two pieces, $\mathcal{H}^0_\pm\left(\nabla_p\right)$ and $\mathcal{H}^1_\pm\left(\nabla_p\right)$. $\mathcal{H}^0_\pm\left(\nabla_p\right)$ contains the eigenfunctions whose eigenvalues may vanish somewhere on P. For λ^+_p an eigenfunction of $L_+\left(p\right)$, $\lambda^+_p \neq 0$ for $p \in P$,

$$L_+\left(p\right)\phi^+_p = \lambda^+_p\,\phi^+_p \tag{11.26}$$

we have that

$$D_+\left(\nabla_p\right)\phi^+_p, \tag{11.27}$$

is an eigenfunction of $L_-\left(p\right)$ with the same eigenvalue, meaning that $D_+\left(\nabla_p\right)$ provides an isomorphism between $\mathcal{H}^1_+\left(\nabla_p\right)$ and $\mathcal{H}^1_-\left(\nabla_p\right)$.

These observations set the stage for the relation between the anomaly and the index theorem. Note that, by definition,

$$D_+\left(\nabla_p\right):\ \mathcal{H}^1_+\left(\nabla_p\right)\ \to\ \mathcal{H}^1_-\left(\nabla_p\right)$$

is invertible. We have already established that $\mathcal{H}^1_+\left(\nabla_p\right)$ and $\mathcal{H}^1_-\left(\nabla_p\right)$ are isomorphic. This means that there is no twisting between $\mathcal{H}^1_+\left(\nabla_p\right)$ and $\mathcal{H}^1_-\left(\nabla_p\right)$. Therefore, any relative twisting between $\mathcal{H}^+\left(p\right)$, $\mathcal{H}^-\left(p\right)$ must originate from $\mathcal{H}^0_\pm\left(\nabla_p\right)$. The finite dimensionality of $\mathcal{H}^0_\pm\left(\nabla_p\right)$ implies that of the vector bundles over P. From now on, let $V_\pm$ denote such bundles. They are easily characterized by their Chern characters, $\mathrm{ch}\left(V_+\right)$ or $\mathrm{ch}\left(V_-\right)$. According to our previous computation of the anomaly, it appears that $\mathrm{ch}\left(V_+\right)$ or $\mathrm{ch}\left(V_-\right)$ are not sufficient to determine the anomaly. The remainder of the chapter is devoted to illustrating this point.

The effective action for a massless Dirac fermion in a $2k$-dimensional Euclidean space is a Gaussian integral:

$$e^{-\Gamma[\nabla,g]} = \int d\psi\,d\bar{\psi}\,\exp\left(-\int d^{2k}x\,\sqrt{g}\,\bar{\psi}\,i\,\displaystyle{\not}D\,\psi\right). \tag{11.28}$$

The classical action is invariant under global chiral rotations of the fermions:

$$\begin{aligned}\psi\ &\to\ e^{i\alpha\bar{\Gamma}}\,\psi\\ \bar{\psi}\ &\to\ \bar{\psi}\,e^{i\alpha\bar{\Gamma}},\end{aligned} \tag{11.29}$$

in part because $\{\bar{\Gamma}, \displaystyle{\not}D\}$ can be shown to be trivial. The effective action in equation (11.28) could be written in terms of the eigenfunctions of $i\,\displaystyle{\not}D$ by expanding $\psi, \overline{\psi}$:

$$\begin{aligned}
\psi(x) &= \sum_k a\,\psi_k(x) \\
\overline{\psi}(x) &= \sum_k b\,\psi_k^+(x) \\
i\,\slashed{D}\,\psi_k &= \lambda_k\,\psi_k.
\end{aligned} \qquad (11.30)$$

And so (11.28) now reads:

$$\int d^{2k}x\,\sqrt{g}\,\overline{\psi}\,i\,\slashed{D}\,\psi = \sum_k \lambda_k\,\overline{b}_k\,a. \qquad (11.31)$$

The measures in formula (11.28) are thus $\prod_k d\overline{b}\,da$, where a and $\overline{b}$ are independent Grassmann numbers. A change of variables

$$\begin{aligned}
\psi &\to \psi + i\alpha(x)\,\overline{\Gamma}\,\psi \\
\overline{\psi} &\to \overline{\psi}i\overline{\Gamma}\,\alpha(x),
\end{aligned} \qquad (11.32)$$

yields a change of the effective action by

$$\int dx\,\overline{\psi}\,i\,\slashed{D}\,\psi \to \int dx\,\overline{\psi}\,i\,\slashed{D}\,\psi + \int d^{2k}x\sqrt{g}\,\alpha(x)\,(\Delta_\mu J_5^\mu); \qquad (11.33)$$

where

$$\begin{aligned}
J_5^\mu &= \overline{\psi}\,\Gamma^\mu\,\overline{\Gamma}\,\psi \\
dx &\equiv d^{2k}x\,\sqrt{g}.
\end{aligned}$$

Ward identities, which imply the conservation of the axial current at the quantum level, arise when expanding (11.33) in powers of $\alpha(x)$, taking into account the invariance of $\Gamma\,[\nabla]$ under changes of variables. The observation is due to Fujikawa [15] that the Jacobian factor appearing in the measure thanks to the change of variables in (11.32) is potentially dangerous. Explicitly, he found that the Jacobian is divergent, and thus the computation of the anomaly reduces to computing the change in the measure. For instance, for small α, the measure is described by the formula:

$$\begin{aligned}
\delta\mathcal{L} &= \exp\left(-2i\int(dx)\,\alpha(x)\sum_k^{i=1}\psi_k^+(x)\,\overline{\Gamma}\,\psi_k(x)\right) \\
&= (\exp -2i\int(dx)\,\alpha(x)\,\Xi(x)).
\end{aligned} \qquad (11.34)$$

The sum is best regulated by a Gaussian cut-off. Using this information allows one to define

$$\begin{aligned}
J &= 2\int dx\,\alpha(x)\sum_k^{i=1}\psi_k^+(x)\,\overline{\Gamma}\,\psi(x) \\
&\equiv 2\int dx\,\alpha(x)\sum_k^{i=1}\psi_k^+(x)\,\overline{\Gamma}\,\psi_k(x)\,e^{-\lambda_k^2/M} \\
&= 2\int(dx)\,\alpha(x)\sum(dx)\,\alpha(x)\sum \psi_k^+(x)\,\overline{\Gamma}\,e^{(-i\slashed{D})^2/M^2}\,\psi_k(x).
\end{aligned}$$

Taking into account the zero momentum Ward identities (that is, the case in which $\alpha(x)$ is close to being a constant), α factorizes out of the integral. Consequently,

$$J = 2 \lim_{M^{2k} \to \infty} \alpha \operatorname{Tr} \overline{\Gamma} e^{(i \not{D})^2/M^2}. \qquad (11.35)$$

When only contributions from the zero sector are considered, we obtain the relation

$$\operatorname{Tr} \overline{\Gamma} \exp \left(-(i \not{D})^2/M^2 = k_+ - k_- = \operatorname{ind} \not{D} \right).$$

The relation then follows that

$$\int (dx)\, \alpha(x)\, \langle \nabla_\mu J_5^\mu \rangle = 2 \int_{M^{2k}} \alpha(x) \left[\hat{A}(M^{2k}) \operatorname{ch}(F) \right]_{\text{vol}} \qquad (11.36)$$

(Compare with (11.4)). Axial anomalies arising in the presence of both gauge and gravitational fields follow as a consequence of formula (11.36). In dimension four, for example, we have

$$\int_{M^{2k}} \langle \nabla_\mu J_5^\mu \rangle = \frac{1}{(2\pi)^2} \int \left(-\operatorname{Tr} F^2 + \frac{r}{24} \operatorname{Tr} R^2 \right). \qquad (11.37)$$

For chiral fermions though, the story is different. For one thing, the effective action cannot be written as

$$\det D_+ = \det i \not{D} P.$$

The reason simply lies in the ill-defined eigenvalue of $D_+ = R_+ \otimes V \to R_- \otimes V$. In principle, looking at the perturbative evolution of the fermionic effective action could solve this issue. Demanding gauge invariance of the effective action under infinitesimal gauge transformation would mean, in this particular instance, that the following formula ought to be satisfied:

$$\Gamma[A - DV] - \Gamma[A] = \int dx\, V^a(x)\, D_\mu \frac{\delta \Gamma[A]}{\delta A_\mu{}^a}. \qquad (11.38)$$

The consistency condition (11.38) holds by virtue of

$$\begin{aligned} e^{-\Gamma[A]} &= \int d\lambda\, d\overline{\lambda} \left(\exp - \int dx\, \overline{\lambda}\, i \not{D}_+ \lambda \right) \\ \frac{\delta \Gamma[A]}{\delta A_\mu{}^a} &= \left\langle \overline{\lambda} \Gamma_\mu P_+ T^a \lambda \right\rangle_A. \end{aligned} \qquad (11.39)$$

The much celebrated statement in field theories, that gauge invariance is equivalent to current conservation, is a mere consequence of formulaes (11.38-39).

According to the material covered in Chapter 10, for complex representations of fermions, the term $\Gamma_\mu [A]$ will generically be anomalous. Hence, the issue before us is to determine whether one can relate the anomaly to more detailed properties of the chiral fermions determinant. Roughly speaking, a theory which does not admit a gauge invariant mass term is potentially anomalous. When such a mass term exists, we could, in principle, regulate the ill-theory by the use of Pauli-Villars fields; this approach is referred to as the Pauli-Villars regularization. Alvarez-Gaumé and Ginsparg [2] have shown a rather simple way to regulate $\Gamma_r [a]$ in terms of how that could de done. The starting point is the basis for the Γ-matrices where $\overline{\Gamma}$ is diagonal. Instead of working with the operator $i \not{D} P_+$, it is useful to now consider the operator

$$\widehat{D} \equiv i\Gamma^\mu \left(\partial_\mu + A_\mu P_+\right). \tag{11.40}$$

This is indeed an elliptic operator acting on Dirac rather than Weyl fermions; so we are still left with an eigenvalue problem. Notice, however, that $\widehat{D}$ is not self-adjoint, so its eigenvalues are complex. This means that one has to simultaneouly take into account left and right eigenfunctions:

$$\begin{aligned}
\widehat{D} \, \phi_k &= \lambda_k \, \phi_k \\
X_k^+ \, \widehat{D} &= \lambda_k \, X_k^+ \\
(X_k, \phi_l) &= \delta_{kl}.
\end{aligned} \tag{11.41}$$

We therefore obtain

$$e^{-\Gamma_r[A]} = \det \widehat{D} \, (A). \tag{11.42}$$

The anomaly is computed using Fujikawa's method [15]. The action,

$$\int dx \, \overline{\psi} \, \widehat{D} \, \psi \tag{11.43}$$

is invariant under the transformation

$$\begin{aligned}
\psi &\to g P_+ \psi + P_- \psi \\
\overline{\psi} &\to \overline{\psi}_- \, g^{-1} + \overline{\psi}_+.
\end{aligned} \tag{11.44}$$

The expansion of ψ, $\overline{\psi}$ is made possible by the terms ϕ_k, X_k^+, and ψ in formula (11.40):

$$\begin{aligned} \psi &= \sum_{i=1}^{k} a\, \phi_k \\ \overline{\psi} &= \sum_{i=1}^{k} \overline{b}\, X_k^+. \end{aligned} \tag{11.45}$$

The measure reads

$$\prod_k d\overline{b}\, d\, a;$$

and this changes the effective action in (11.42) in such a way that it is now

$$\sum_{k}^{i=1} \lambda_k \overline{\jmath}_k\, i_k.$$

Moreover, equation (11.41) is easily extracted after a Gaussian integration.

The Jacobian factor arising under infinitesimal gauge transformation is

$$\begin{aligned} \delta V\, \Gamma &= \int V^i\, D_\mu \frac{\delta\Gamma[A]}{\delta A_\mu{}^a} \\ &= \int_{2k}^{0} dx\, \mathrm{Tr}\, v\,(x)\, \overline{\Gamma}\, e^{-(i\widehat{D})^2/M^2}\, \delta\,(x-y). \end{aligned} \tag{11.46}$$

Its 4-dimensional incarnation is

$$\begin{aligned} \delta_V\, \Gamma\,[A] &= \tfrac{1}{24\pi^2} \int d^4x\, \mathrm{Tr}\, v\, \epsilon^{\lambda\mu\alpha\beta}\, \partial_\lambda \left(A_\mu\, \partial_\alpha A_\beta + \tfrac{1}{2} A_\mu A_\alpha A_\beta \right) \\ &= \tfrac{1}{24\pi^2} \int \mathrm{Tr}\, V\, d \left(A\, dA + \tfrac{1}{2} A^3 \right). \end{aligned} \tag{11.47}$$

The leading term in equation (11.46) is given by

$$\frac{i^{k+2}}{(2\pi)^k\,(k+1)!}\, \mathrm{Tr}\, \left[V\,(dA)^k \right]. \tag{11.48}$$

We have reached the conclusion that the term $\Gamma_r\,[A]$ can indeed be defined by $\det \widehat{D}\,(A)$.

The next order of priority for us is to demonstrate that the form of the anomaly (11.46) is a consequence of the index theorem (or otherwise, a consequence of the index theorem for families of operators). In order to state the relationship between the $2k$-dimensional anomaly and the $2k+2$ index theorem, we begin by considering a one-parameter family of gauge transformations:

$$\begin{aligned} g\,(\theta,x) &= S^1 \times S^{2k} \to G \\ g\,(0,x) &= g\,(2\pi,x) = 1. \end{aligned} \tag{11.49}$$

The classification of these maps is given by $\pi_{2k+1} G$ and extends to the cases for which $\pi_1 G = 0$ and $\pi_{2k} G = 0$. The equivalence transformations in (11.49) yield a one-parameter family of gauge field configurations which reads

$$A^\theta = g^{-1} (A + d) g. \tag{11.50}$$

To go from A^θ to say, $A^{\theta + \delta\theta}$, is the result of an infinitesimal gauge transformation (with gauge parameter $g^{-1} \delta_\theta g$):

$$\begin{aligned} A^{\theta + \delta\theta} &= A^\theta + \delta\theta\, D_{A^\theta} (g^{-1} \partial_\theta g) \\ D_{A^\theta} v &= dv + \left[A^\theta, v \right]. \end{aligned} \tag{11.51}$$

The anomaly obtained by the infinitesimal transformation (11.49) is

$$- \delta_\theta \Gamma \left[A^\theta \right] = i \frac{d\omega (A, \theta)}{d\theta}. \tag{11.52}$$

The term $e^{i\omega (A, \omega)}$ is a function mapping $S^1 \to S^2$. As such, the anomaly measures the local winding number of $e^{i\omega (A, \omega)}$. To obtain the total winding number, one has to extract the integrated anomaly along the one-parameter family of gauge transformations shown in equation (11.49):

$$- \int_0^{2\pi} d\theta \frac{d\Gamma \left[A^\theta \right]}{d\theta} = i \int_0^{2\pi} d\theta \frac{d\omega(A, \theta)}{d\theta} = 2\pi i; \tag{11.53}$$

here n stands for the winding number of $\exp i\omega (A, \theta)$.

In view of this, we need to find an appropriate $(2k + 2)$ dimensional Dirac operator whose index equals the winding number (11.51). The integral form for the index, we shall learn, will then generate a term for the anomaly in the form (11.46). The fermion determinant defines a complex line bundle over the 2-sphere whose transition function is the imaginary part of the effective action, $\exp i\omega (A, \theta)$. In turn, the bundles are classified by the winding number of the transition function (11.51). The issue before us, at this point, is to compute a representative of the first Chern class. This is equivalent to computing a particular form of the anomaly.

In $2k + 2$ dimensions, the Dirac operator takes the form

$$i \not{D}_{2k+2} = i \sum_{i=1}^{2k+2} (\partial_i + A_i) \overline{\Gamma}^a. \tag{11.54}$$

The use of the index theorem (11.4) yields

$$\text{ind}\, i\, \not{D}_{2k+2} = \frac{i^{k+1}}{(2\pi)^{k+1}\,(k+1)!} \int_{S^2 \times S^{2k}} \text{Tr}\, \tilde{F}^{k+1}. \qquad (11.55)$$

F is the $(2k + 2)$-dimensional gauge field strength:

$$\tilde{F} = (d_t + d_\theta + d)\, A + A^2.$$

Formula (11.53) takes on the final form

$$\text{ind}\, i\, \not{D}_{2k+2} = \frac{(-1)^k\, i^{k+1}}{(2\pi)^{k+1}} \frac{k!}{(2k+1)!} \\ \times \int_{S^1 \times S^{2k}} \text{Tr}\, (g^{-1}\, dg)^{2k+1}. \qquad (11.56)$$

The resulting formula

$$\begin{aligned} &- \int d\theta\, \frac{d\Gamma\left[A^\theta\right]}{d\theta} \\ &= 2\pi\, \frac{1}{(2\pi)^3\, 2!} \int_{S^1 \times S^4} \text{Tr}\, \tilde{F}^3 \\ &= \frac{1}{24\pi^2} \int d\partial \int_{S^4} \text{Tr}\, g^{-1}\, \partial_\theta g\, d\, (A^\theta\, dA^\theta + \tfrac{1}{2}\, A^{\theta^3}) \end{aligned} \qquad (11.57)$$

agrees term by term with (11.46).

We now pause to discuss the assertion made a while ago that the Chern character $\text{ch}\,(V_\pm)$ of vector bundles on M^{2k} is not relevant in determining the anomaly. In keeping with the spirit of our discussion above, the anomaly is generated by the relative twisting of $\psi_+^0\,(t, \theta)$ and $\psi_-^0\,(t, \theta)$, the eigenfunctions of $\not{D}^{(t,\theta)}$ which are trivial at (t_0, θ_0). The winding number obtained earlier measures the twisting of ψ_- and of course, ψ_+ around (t_0, θ_0). Topologically, this amounts to determining the first Chern class of a line bundle. To quantify the relative twisting of V_- and V_+ involves the introduction of the difference bundle $V_+ \ominus V_-$. From Chapters 6 and 7, we know how to define the direct sum of vector bundles $E \oplus F$. The very definition of Chern classes gives $\text{ch}\,(E \oplus F) = \text{ch}\,(E) + \text{ch}\,(F)$ and $c\,(E \oplus F) = c\,(E) \cdot c\,(F)$. The index is therefore

$$\text{ind}\, D = \dim\,(\ker D_-) - \dim\,(\ker D_+); \qquad (11.58)$$

where

$$\text{ch}\,(\text{ind}\, D) = \int_{M^{2k}} \hat{A}\,(B)\, \text{ch}\,(V). \qquad (11.59)$$

We are recall that $\hat{A}(B)$ is none other than the Dirac genus originally given by formulae (11.4) and (11.5). Furthermore, $B = P \times M^{2k}$.

For gravitational anomalies, the charactersistic polynomial in (11.57) should be interpreted as a polynomial in the Pontrjagin classes. Since Pontrjagin classes are forms of degree divisible by four, the relation follows:

$$\mathrm{ch}(\mathrm{ind}\,D) = c(\mathrm{ind}\,D).$$

The dimension in which it does not vanish is $d = 4k + 2$ (in this case B has dimension $4k$) [8]. It is precisely because of this fact that gravitational anomalies exist only in dimensions $4k + 2$.

In the presence of a gravitational field, a theory describing chiral fermions is consistent when invariance of the effective action under both diffeormorphisms and local Lorentz transformations is achieved. In view of this, for the characteristic polynomial to generate anomalies, a non-trivial index theorem for a two-parameter family of Dirac operators ought to exist. The latter condition is required in part because the anomaly cancellation conditions are the same for diffeomorphism or Lorentz anomalies: one can shift the anomaly from diffeormorphism to local Lorentz transformations simply by adding a local counterterm to the theory's effective action.

How does one go about finding a two-parameter family of Dirac operators with a non-trivial family index theorem in dimensions $4k + 2$? The answer essentially lies with the characteristic polynomial (11.57) which, as we have seen, gives the anomalies. The discovery, by Atiyah [16], of two-parameter families of two-dimensional Dirac operators with a non-trivial index, certainly provides us an adequate framework from which a detailed analysis of the gravitational anomaly in (11.57) could be carried out.

Using Atiyah's result, we see for instance, that for spin $1/2$, $V = 0$, $\mathrm{ch}(V) = 1$ (and $\hat{A}(B)$ defined as in (11.4)):

$$I_{1/2} = \prod \frac{x_i/2}{\sinh x_i/2}. \tag{11.60}$$

At this point, we need only expand $I_{1/2}$ to order $2k$ to exhibit the $4k$-form characterizing the gravitational anomaly.

11.4   References

[1] Atiyah, M. F. and Singer, I.: *Dirac Operators Coupled to Vector Potentials*, Proc. Nat. Academy Sciences USA 81 (1984) 2597.

[2] Alvarez-Gaumé, L. and Ginsparg, P.: *The Topological Meaning of Non-Abelian Anomalies*, Nucl. Phys. B243 (1984) 449.

[3] Moore, G. and Nelson, P.: *The Aetiology of Sigma Model Anomalies*, Commun. Math. Physics 100 (1985) 83.

[4] Bardeen, W.: *Anomalous Ward Identities in Spinor Field Theories*, Phys. Reviews 184 (1969) 1848.

[5] Alvarez-Gaumé, L.: *An Introduction to Anomalies*, in **Fundamental Problems of Gauge Field Theory**, Eds. Velo, G. and Wightman, A. NATO-ASI Series B, Vol. 141 (1985) 93-206.

[6] Eguchi, T.; Gilkey, P. B. and Hanson, A. I.: **Gravitation, Gauge Theories and Differential Geometry**, Physics Reports 66 No. 6 (1980) 213-393.

[7] Atiyah, M. F. and Singer, I.: Annals of Mathematics 87 (1968) 485-546. – idem, 93 1 (1971) 119-139.

[8] Alvarez-Gaumé, L. and Witten, E.: *Gravitational Anomalies*, Nucl. Physics B 234 (1983) 269.

[9] Chern, S. S.: **Complex Manifolds Without Potential Theory**, Van Nostrand 1967.

Milnor, J. and Strasheff, J.: **The Theory of Characteristic Classes**, Princeton University Press 1974.

[10] Alvarez-Gaumé, L.; Della Pietra, S., and Moore, G.: *Anomalies and Odd Dimensions*, Annals of Physics 163 (1985) 288.

[11] Niemi, A. and Semenoff, G.: Phys. Rev. Letters 55 (1985) 927.

[12] Witten, E.: *Global Gravitational Anomalies*, Commun. Math. Phys. 100 (1985) 197-229.

[13] Freed, D. S.: *Determinants, Torsion and Strings*, Commun. Math. Physics 107 (1986) 483-513.

[14] Freed, D. S.: $\mathbb{Z}/k$-*Manifolds and Families of Dirac Operators*, Invent.

Mathematica 92 (1988) 243-254.

[15 Fujikawa, K.: Phys. Rev. Letters 42 (1979) 1195.

– idem, 44 (1980) 1733.

Phys. Reviews D 21 (1980) 2848.

– idem, D 22 (1980) 1499 (E).

– idem D 23 (1981) 2262.

– idem D 29 (1984) 285.

[16] Atiyah, M. F.: *The Signature of Fiber Bundles*, in **Collected Mathematical Papers in Honor of K. Kodaira**, Tokyo University Press 1969.

Chapter 12

Global Anomalies

We are (finally) in the position to present an exhaustive account of global anomalies. Global anomalies occur when large gauge or diffeomorphim transformations fail to be symmetries of the corresponding quantum theory. There are, to date, two classes of global anomalies: the global gauge anomaly and the global gravitational anomaly. Global gauge anomalies arise whenever a theory's effective action is not invariant under large gauge transformations. The extent to which *large* takes on a pathological meaning relates to whether the effective action is well-defined under small gauge transformations. These are usually detected by anomalies of the perturbative or chiral type. When the effective action $\mathcal{L}_{\text{eff}}$ is well-defined, large gauge transformations of a global character change it in a highly non-trivial way. Global gravitational anomalies, on the other hand, reflect the lack of invariance of the effective action under the group of equivalence classes of diffeomorphisms that *cannot* be smoothly deformed to the identity, the so-called mapping class groups.

12.1   An SU (2) Global Gauge Anomaly

Elements of homotopy theory are necessary to comprehend this class of anomalies. Primarily, the use of homotopy groups is required. These groups are easily catalogued by Bott's periodicity theorem for $i < 2n$ [1] (with G a given gauge group:)

$$\begin{cases} \pi_i\left(G\left(n\right)\right) = \mathbb{Z} & \text{for } i = \text{odd} \\ \pi_i\left(G\left(n\right)\right) = 0 & \text{for } i = \text{even} \end{cases}. \tag{12.1}$$

It is a periodicity in the sense that for $n = \infty$, $\pi_i\left(G\left(\infty\right)\right) = \pi_{i+2}\left(G\left(\infty\right)\right)$ with $\pi_0\left(G\left(\infty\right)\right) = 0$ and $\pi_1\left(G\left(\infty\right)\right) = \mathbb{Z}$. Global gauge anomalies rely on Bott periodicity and on non-simply connected gauge groups, say $G\left(n\right)$ (in the latter, we take account of loops which are not homotopic to the identity map; these are the relevant ones for evaluating the pathologies). The best available illustration of such is Witten's SU (2) global anomaly [2], which we proceed to succinctly analyze below.

The problem is essentially easy to formulate conceptually, but, nonetheless requires a good deal of technical knowledge to solve. In dimension four, the number of Weyl doublets is perturbatively anomaly-free. However, as pointed out by Witten in reference [2], the theory is mathematically inconsistent because, according to Bott periodicity, the gauge group of interest, G, is not simply connected and therefore does admit loops which are not homotopic to the identity map. When one encounters such a situation, global anomalies are bound to arise, and thus need to be cancelled if the theory is to be consistent. Let us make this a bit more explicit. Let $G = \text{SU}(2)$ be such a group. The formula

$$\pi_4\left(\text{SU}(2)\right) = \mathbb{Z}_2, \tag{12.2}$$

suggests that, in four dimensional Euclidean space, there is a gauge transformation, $T\left(x\right) \to 1$ as $|x| \to \infty$. For $\pi_n\left(\text{SU}(n)\right) = \mathbb{Z}$, it follows that the transformation in question *cannot* be continuously deformed to the identity. However, for $\mathbb{Z}_2$ homotopy groups, the gauge transformation wraps twice around SU (2) and thus can be deformed to the identity. Now, in the absence of fermions, the Euclidean path integral is of the form:

$$\int \left(d\,A_\mu\right) \exp\left(-\frac{2}{2g^2} \int d^4\,x \operatorname{tr} F_{\mu\nu}\,F^{\mu\nu}\right). \tag{12.3}$$

To a given gauge field configuration A_μ corresponds a conjugate field,

$$A_\mu^T = T^{-1}\,A_\mu\,T - i\,T^{-1}\,\partial_\mu T,$$

which makes the same contribution to the functional integral (12.3). Hence, to each contribution originating from the gauge field is to be associated another contribution. Such double counting is harmless in the absence of fermions as it cancels out when computing the vacuum expectation value.

We now introduce fermions. The partition function of the Euclidean version of the model of a doublet of massless fermions coupled to an SU(2) gauge field is

$$Z = \int d\psi_L \, d\psi_L^+ \int dA_\mu \exp \\ \left(-\int d^4 x \, \left[(1/2g^2) \operatorname{tr} F_{\mu\nu}^2 + \psi_L^+ \, i \, \rlap{/}{D} \, \psi_L\right]\right). \tag{12.4}$$

A_μ is the SU(2) gauge field; ψ_L a left-handed Weyl fermion doublet; g the gauge coupling constant, and $\rlap{/}{D} = D_\mu \gamma_\mu$ is the Dirac operator restricted to act on a Weyl doublet. The fermionic part of the integral (12.4) is ill-defined. It can be formally integrated as the square root of functional integrals over Dirac fermions. As such, it implies the doubling of the fermionic degrees of freedom, meaning we need to consider two Weyl-handed doublets instead of one. Note that the 1/2 representation of SU (2) is pseudoreal. Consequently, a left-handed Weyl doublet can be mapped to a right-handed one. A theory with two left-handed Weyl doublets is therefore equivalent to a vector-like one with a single Dirac doublet. The fermionic functional integral is obtained through det $(i \, \rlap{/}{D})$; it is well defined. The formula then follows:

$$\int d\psi_L \, d\psi_L^+ \exp \int \psi_L^+ \, i \, \rlap{/}{D} \, \psi_L = (\det i \, \rlap{/}{D})^{1/2}. \tag{12.5}$$

At this point we run into a problem: the sign of the square root is ill-defined. It is precisely this ambiguity that defined the global gauge anomaly problem. For a given gauge field, A_μ, the sign of $(\det i \, \rlap{/}{D})^{1/2}$ can be determined arbitrarily. When certain consistency conditions, such as the Schwinger-Dyson equations, are taken into consideration though, it turns out that the fermion integral $(\det i \, \rlap{/}{D})^{1/2}$ ought to be defined in such a way that it varies smoothly as A_μ is varied. This procedure guarantees the fermion integral to be invariant under infinitesimal gauge transformations, that is to say, the sign does not change abruptly. Such gauge invariance is, however, not necessarily guaranteed under the *large*, topologically, non-trivial gauge transformation T. In reference [2], it is argued that $(\det i \, \rlap{/}{D})^{1/2}$ is indeed odd under T. Here is how.

Roughly, the lack of invariance under T is best illustrated by the formula

$$\det i \not{D} \left(A_\mu\right)^{1/2} = -\left(\det i \not{D} \left(A_\mu^T\right)\right)^{1/2}. \tag{12.6}$$

Formula (12.6) is the statement that a continuous variation of A_μ to A_μ^T generates the opposite sign of the square root.

How does one solve the issue of the ill-defined sign of the square root? A clever solution, proposed in [2], is to first write down the root in (12.5) as the product of all positive eigenvalues of a Dirac operator, and then continue analytically. As a result of this procedure, the partition function,

$$\begin{aligned} Z \;=\; &\int dA_\mu \left(\det i \not{D}\right)^{1/2} \\ &\exp\left(-(1/2g^2) \int d^4x \operatorname{tr} F_{\mu\nu}^2\right), \end{aligned} \tag{12.7}$$

vanishes due to the contribution of opposite signs from A_μ and A_μ^T. This should come as no surprise since, as we are continuously varying the external field value from A_μ to A_μ^T, an odd number of these eigenvalues flows through zero, switching their sign in the process. This very phenomenon is related to the existence of an odd number of normalizable zero modes for a well-defined five-dimensional Dirac operator, $\not{D}_5$, in a topologically non-trivial gauge field. It is noted in [2] that $\not{D}_5$, for an SU(2) doublet, is a real, antisymmetric operator. Its eigenvalues either vanish or are imaginary and occur in complex conjugate pairs. A variation of the gauge field A_μ implies that the number of zero eigenvalues of $\not{D}_5$ changes only if a complex conjugate pair of eigenvalues moves to, or away from, the origin. The Atiyah-Singer theorem [3] requires that $\not{D}_5$ possess an odd number of zero eigenvalues. This number, modulo two, corresponds to the mod two index of the Dirac operator. Having determined this, it then follows that $(\det (i \not{D}))^{1/2}$ is odd under the topologically non-trivial gauge transformation T.

12.2   Global Gravitational Anomalies

Our analysis of global gravitational anomalies shall begin with the $N = 2$, $D = 10$ dimensional perturbative anomaly-free chiral supergravity theory. Then, in a second phase, we will address the question of whether the $E_8 \times E_8$ heterotic superstring theory is global anomaly-free. There will be, moreover, a third focus on the manifestation and cancellation of global gravitational anomalies in some six-dimensional supergravity theories. In the next chapter, we will discuss the relationship between mapping class groups and global gravitational anomalies in dimension three particularly as they relate to topological quantum field theories and Chern-Simons-Witten theories. Any serious study of global gravitational anomalies in ten dimensions, it should be said, requires a specific knowledge of homotopy (i.e. exotic) spheres.

12.2.1   A Case Study of the $N = 2$, $D = 10$ Supergravity Theory

This theory is known to be chiral anomaly-free [4]. The foremost investigation of global gravitational anomalies in this theory is that of reference [5]. Consider a ten-dimensional manifold, M^{10}, in which the theory is defined; let the group of diffeomorphisms of this manifold be Diff (M). We are interested in diffeomorphisms which become the identity at infinity, and this suggests we first determine how many components of Diff (M) there are in dimension ten. Finding the number of components of this group is actually a highly non-trivial exercise since, to begin with, there is no general method that can offer a satisfactory answer. So, in order to tackle this problem, one further considers a special class of diffeomorphisms which are different from the identity when restricted to some ten-dimensional ball B^{10}, and become the identity outside. The appropriate use of algebraic topology techniques shows that these diffeomorphisms are the same as diffeomorphisms of S^{10}, that is,

$$\text{Diff}\left(B^{10}\right) \simeq \text{Diff}\left(S^{10}\right). \tag{12.8}$$

Formula (12.8) requires a thorough evaluation of π_0 (Diff (S^{10})), the mapping class group of S^{10}. To detect the gravitational anomaly is now equivalent to asking whether the $N = 2$, $D = 10$ supergravity theory considered is invariant under the action of these disconnected diffeomorphism components. The answer to this question depends, in turn, on our evaluation of Milnor-Kervaire 11-dimensional exotic spheres [6]. Milnor has shown that exotic $(n + 1)$-spheres and topological classes of diffeomorphisms of the (standard) n-sphere are in one to one correspondence.

For a twelve-manifold without boundary, M^{12}, the integral of products of

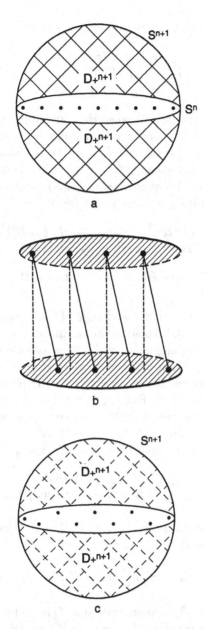

Figure 12.1: Sketched here is the topological construction of exotic spheres obtained via surgery on the standard n-sphere.

characteristic classes which are relevant as topological invariants, are:

$$I_1 = \int_{M^{12}} \left(\operatorname{Tr} R^2\right)^3,$$
$$I_2 = \int_{M^{12}} \operatorname{Tr} R^2 \cdot \operatorname{Tr} R^4, \tag{12.9}$$
$$I_3 = \int_{M^{12}} \operatorname{Tr} R^6.$$

These quantities are invariant because they are independent of the metric and the connection. When M^{12} has a non-empty boundary, however, I_1, I_2 and I_3 lose their topological invariance. On the other hand, if, on the boundary of M^{12} one can solve $\operatorname{Tr} R^2 = dH$, then there is no obstruction to generalizing some of the quantities in (12.9) to a twelve-manifold with boundary. The generalization takes on the form

$$\bar{I}_1 = \int_{M^{12}} \left(\operatorname{Tr} R^2\right)^3 - \int_{\partial M^{12} = \Sigma^{11}} H \left(\operatorname{Tr} R^2\right)^2,$$
$$\bar{I}_2 = \int_{M^{12}} \operatorname{Tr} R^2 \cdot \operatorname{Tr} R^4 - \int_{\Sigma^{11}} H \operatorname{Tr} R^4. \tag{12.10}$$

From here on, we will take $\partial M^{12} = \Sigma^{11}$ to be the Milnor-Kervaire 11-dimensional exotic sphere.

Are $\bar{I}_1$ and $\bar{I}_2$ invariant under a change of metric on M^{12}? The answer to this, as explained in [5], involves some technicalities. The procedure used to obtain generalized topological invariants is as follows. First, we write $\delta \operatorname{Tr} R^2 = d\Lambda$ for a variation of $\operatorname{Tr} R^2$ under a change of metric. It follows that $\delta H = \Lambda$ does preserve the requirement that $\operatorname{Tr} R^2 = dH$ on ∂M^{12}, and consequently $\delta \bar{I}_1 = \delta \bar{I}_2 = 0$. But (12.10) is still not well defined via these arguments. For this argument to hold, one needs to establish that $\bar{I}_1$ and $\bar{I}_2$ are independent of the choice of H for $dH = \operatorname{Tr} R^2$. Let $dH = dH' = \operatorname{Tr} R^2$ (i.e. $d(H - H') = 0$).) In going from H to H', the quantity $\bar{I}_1$ changes by

$$\Delta \bar{I}_1 = \int_{\Sigma^{11}} (H - H') \left(\operatorname{Tr} R^2\right)^2$$
$$= \int_{\Sigma^{11}} (H - H') \cdot dH \cdot \operatorname{Tr} R^2 = 0. \tag{12.11}$$

Formula (12.11) is essentially the statement that $\bar{I}_1$ is well-defined. Looking now at $\bar{I}_2$, its variation is

$$\Delta \bar{I}_2 = \int_{\Sigma^{11}} (H - H') \operatorname{Tr} R^4, \tag{12.12}$$

and does not vanish [5]. To demonstrate that $\bar{I}_2$ is well-defined on the boundary $\partial M^{12} = \Sigma^{11}$, we consider the term $\operatorname{Tr} R^4 = dB$. The variation of $\bar{I}_2$

now becomes:

$$\Delta \bar{I}_2 = \int_{\partial M^{12}} (H - H') \, dB$$
$$= - \int_{\Sigma^{11}} d(H - H') \cdot B = 0. \qquad (12.13)$$

To resume what we have done so far: I_1 can be generalized to a topological invariant of M^{12} with boundary, provided that $\text{Tr}\, R^2 = dH$ on Σ^{11}. I_2 can similarly be generalized if $\text{Tr}\, R^2 = dH$ and $\text{Tr}\, R^4 = dB$ on Σ^{11}. I_3, on the other hand, does not admit a similar generalization.

Our next focus is to concern ourselves with the seven-and eleven-dimensional cases. A closed eight-dimensional manifold, M^8, has the following expressions for its Pontrjagin numbers:

$$p_1^2 = \frac{1}{(2\pi)^4} \int_{M^8} \frac{1}{4} \left(\text{Tr}\, R^2\right)^2 ,$$
$$p_2 = \frac{1}{(2\pi)^4} \int_{M^8} \left(\frac{1}{8} \left(\text{Tr}\, R^2\right)^2 - \frac{1}{4} \text{Tr}\, R^4\right). \qquad (12.14)$$

According to the discussion above, p_2 has no known generalization for 8-manifolds with boundary. The quantity p_1^2, however, admits a generalization to a manifold with boundary, and on the boundary one would expect $\text{Tr}\, R^2 = dH$ to be exactly solvable. The index of the Dirac operator for an 8-dimensional spin manifold is of the form

$$\text{ind}\, i\,\slashed{D} = \frac{1}{5760} (7p_1^2 - 4p_2). \qquad (12.15)$$

The signature reads

$$\sigma = -\frac{1}{45} p_1^2 + \frac{7}{45} p_2. \qquad (12.16)$$

Combining (12.15-16) yields

$$\text{ind}\, i\,\slashed{D} = \frac{1}{896} (p_1^2 - 4\sigma). \qquad (12.17)$$

Since M^8 is a compact manifold, it follows that the right hand side of (12.17) is an integer.

On the basis of formula (12.17), we define the topological invariant

$$\lambda(\Sigma^7) = \frac{1}{896} \left(p_1^2(M^8) - 4\sigma(M^8)\right) \bmod 1, \qquad (12.18)$$

where Σ^7 is the boundary of the oriented eight-dimensional spin manifold. This invariant depends only on Σ^7. To prove this, take two manifolds, M and M', with boundary Σ. Glue them together while preserving the orientation $X = M + (-M')$. The resulting manifold has no boundary. Hence,

$$\lambda(M) - \lambda(M') = \text{ind}\, \not{D}(X) = \mathbb{Z}.$$

Thus $\lambda(\Sigma)$ depends only on the manifold Σ. The case in which Σ is the 7-exotic sphere implies that M has signature equal to 8 (i.e. $\sigma(M) = 8$), and $p_1 = 0$. With these descriptions in hand, one deduces that $\lambda(\Sigma^7) = -\frac{1}{28}$. But $\lambda = 0$ for the standard sphere S^7, therefore Σ^7 must be an exotic sphere. The connected sum of Σ^7 generates exactly twenty-seven exotic spheres.

Having studied the toy model Σ^7, let us move now to the eleven-dimensional case. In a given closed twelve-manifold M^{12}, the relevant quantities are:

$$
\begin{aligned}
p_1^3 &= \tfrac{1}{(2\pi)^6} \int \left[\tfrac{1}{8} \left(\text{Tr}\, R^2 \right)^3 \right], \\
p_1 p_2 &= \tfrac{1}{(2\pi)^6} \int \left[\tfrac{1}{8} \left(\text{Tr}\, R^2 \right) \left(\text{Tr}\, R^4 \right) - \tfrac{1}{16} \left(\text{Tr}\, R^2 \right)^3 \right], \\
p_3 &= \tfrac{1}{(2\pi)^6} \int \left[-\tfrac{1}{6} \text{Tr}\, R^6 + \tfrac{1}{8} \text{Tr}\, R^2 \, \text{Tr}\, R^4 - \tfrac{1}{48} \left(\text{Tr}\, R^2 \right)^3 \right],
\end{aligned}
\tag{12.19}
$$

together with the index:

$$
\text{ind}\, i\, \not{D} = \frac{1}{2^6} \left(-\frac{31}{15120} p_1^3 + \frac{11}{3780} p_1 p_2 - \frac{1}{945} p_3 \right),
\tag{12.20}
$$

and the signature:

$$
\sigma = \left(\frac{2}{945} p_1^3 - \frac{13}{945} p_1 p_2 + \frac{62}{945} p_3 \right).
\tag{12.21}
$$

It is best to determine $\frac{1}{2}$ the Dirac index. This is achieved by a judicious use of equations (12.19), (12.20) and (12.21); the result is:

$$
\frac{1}{2} \text{ind}\, i\, \not{D} = \alpha p_1^3 + \beta p_1 p_2 - \frac{\sigma}{8 \cdot 992};
\tag{12.22}
$$

where α, and β are rational numbers. Witten's computations [5] tell us that (12.22) is always an integer in twelve dimensions. As a consequence, the term $\alpha p_1^3 + \beta p_1 p_2 - \frac{\sigma}{8 \cdot 992}$ is an integer for closed twelve-dimensional spin manifolds.

Now, consider the 11-dimensional exotic sphere, Σ^{11}, with $\operatorname{Tr} R^2 = dH$, and $\operatorname{Tr} R^4 = dK$. Taking Σ^{11} to be the boundary of the twelve-manifold M^{12} gives rise to the invariant:

$$\lambda\left(\Sigma^{11}\right) = \alpha p_1^3 \left(M^{12}\right) + \beta p_1 p - 2 \left(M^{12}\right) - \frac{\sigma(M)}{8 \cdot 992}. \qquad (12.23)$$

As before, modulo one, the invariant (12.23) depends only on the Σ^{11} and *not* on any given choice of M^{12}. Otherwise, λ would be an integer if Σ^{11} was a standard sphere, which is a contradiction in terms. The example is due to Milnor [6], of an exotic sphere Σ^{11} bounding M^{12}, having the property that

$$\lambda\left(\Sigma^{11}\right) = -\frac{1}{992}, \qquad (12.24)$$

($\sigma\left(M^{12}\right) = 8$ in this case.) From (12.24) we have a family of 991 exotic spheres via connected sums of the Milnor sphere Σ^{11}. A statement to this effect is encoded in Milnor's observation [3] that exotic 11-spheres bound twelve-manifolds whose signature is divisible by eight. Witten's insight was the realization that a general formula for global gravitational anomalies could be derived in terms of the variation of the effective action that takes mostly into account the Hirzebruch signature and Milnor's observation. Under large diffeomorphism transformations, Witten noticed, the action changes by

$$\Delta \mathcal{L}_{\text{eff}} = 2\pi i \frac{\sigma\left(M^{12}\right)}{8},$$

and the theory is global gravitational anomaly-free when

$$\Delta \mathcal{L}_{\text{eff}} = 0 \bmod 2\pi i.$$

We will arrive att this formula in a short while.

For now, let us give the flavor of the machinery necessary to carry out the computations in the case of $N = 2$, $D = 10$ chiral supergravity theory.

We have already shown, namely in Chapters 10 and 11, that this theory is perturbative and chiral anomaly-free [4]. In chapter 11, the anomalies cancellation, it was noted, is function of the Hirzebruch signature $\sigma(M)$ and the index of the Rarita-Schwinger and Dirac operators in twelve dimensions:

$$\frac{\sigma}{8} = \operatorname{ind}(R - S) - 3 \operatorname{ind} i \, \slashed{D}. \qquad (12.25)$$

In ten-dimensional Minkowski space, optimal results are achieved by the inclusion of Weyl-Majorana spinors. The effective action, $\mathcal{L}_{\text{eff}}$ for such a spinor is half of that $\mathcal{L}'_{\text{eff}}$ for a Weyl spinor. We want to compute the difference

$$\mathcal{L}'_{\text{eff}}\left[g^{\pi}_{\mu\nu}\right] - \mathcal{L}_{\text{eff}}\left[g_{\mu\nu}\right],$$

where $g_{\mu\nu}$ is the metric and π a diffeormorphism of M. The effective action is written as a functional of the metric. In order to carry out the computation, we choose a one-parameter family of metrics $g^t_{\mu\nu}$ which interpolates between $g_{\mu\nu}$ and $g^{\pi}_{\mu\nu}$. The anomaly affects only the imaginary part of $\mathcal{L}_{\text{eff}}$, and therefore, what is to be computed is simply:

$$\Delta \mathcal{L}'_{\text{eff}} \equiv i \int_1^0 dt\, \frac{d}{dt} \operatorname{Im} \mathcal{L}_{\text{eff}}\left[g^t_{\mu\nu}\right]. \tag{12.26}$$

A series of arguments are then applied [5] to give

$$\mathcal{L}'_{\text{eff}} = i\pi\eta(0); \tag{12.27}$$

here $\eta(0)$ stands for the spectral asymmetry of the Dirac operator evaluated on the manifold $(M \times S^1)_{\pi}$, a cylinder with base M siding the interval $[0, 1]$, whose top and bottom are identified with the twist automorphism π.

Topology-wise, what we have done is to fiber the manifold M over the circle S^1 with transition function equal to π. When taking N_D Weyl-Majorana fermions into account, the formula (12.27) becomes:

$$\Delta \mathcal{L}'_{\text{eff}} = N_D \frac{i\pi}{2} \eta(0). \tag{12.28}$$

A similar result applies to the Rarita-Schwinger field. It is important though, to make use of the fact that in dimensions $n + 1$, a Rarita-Schwinger field decomposes to a Rarita-Schwinger field plus a spinor in its n-dimension component. We also take into account the Weyl-Majorana condition. Under this, the change in the effective action is

$$\Delta \mathcal{L}''_{\text{eff}} = N_R \frac{i\pi}{2} \left(\eta_R(0) - \eta_D(0)\right). \tag{12.29}$$

In this formula, N_R is taken to represent the number of spin 3/2 Weyl-Majorana fields while η factors in the ghosts contribution. Overall, for a theory with N_D, N_R and N_S chiral Dirac, Rarita-Schwinger, and self-dual tensor fields, there is a general formula for the effective action change under a diffeomorphism π, namely

$$\Delta \mathcal{L}_{\text{eff}} = \frac{i\pi}{2} \left(N_D \, \eta_D + N_R \left(\eta_R - \eta_D \right) - \frac{N_S}{2} \eta_S \right). \qquad (12.30)$$

The inclusion of self-dual tensor fields requires a bit of attention. In particular, the change in the effective action in the interpolation from $g_{\mu\nu}$ to $g_{\mu\nu}^\pi$ spells out the need for the spectral asymmetry of specified operators. These operators are $\star d$-forms acting on even forms on $(M \times S^1)_\pi$. Note that $\star$ is a Hodge operator:

$$\star : \mathbb{C}^\infty \left(\Lambda^p \right) \to \mathbb{C}^\infty \left(\Lambda^{n-p} \right).$$

With the inclusion of self-dual tensor field N_S, the variation in the effective action becomes

$$\Delta \mathcal{L}_{\text{eff}} = -\frac{i\pi}{4} \eta \left(0 \right); \qquad (12.31)$$

the minus sign originating from the different statistics with respect to the fermions. Applying the Atiyah-Patodi-Singer Index theorem to (12.30) gives

$$\begin{aligned}
\tfrac{1}{2} \eta_D &= -\operatorname{ind} \slashed{D} + \int_{M^{12}} \widehat{A} (M^{12}); \\
\tfrac{1}{2} \eta_R &= -\operatorname{ind} (R - S) + \operatorname{ind} i \slashed{D} + \int_{M^{12}} \left[K (M^{12}) - \widehat{A} (M^{12}) \right]; \quad (12.32) \\
\eta_S &= -\sigma (M^{12}) + \int_{M^{12}} L (R).
\end{aligned}$$

Some definitions are in order: $\operatorname{ind} \slashed{D}$ and $\operatorname{ind} R$ are the Dirac and Rarita-Schwinger indices on M^{12}, and σ is the signature of M^{12}. As for the terms $\widehat{A}$, K, and L, they are polynomials in the curvature tensor, R, whose integral over M^{12} yields precisely $\operatorname{ind} \slashed{D}$, $\operatorname{ind} R$, and σ. K, in particular, is the characteristic polynomial for the index of the Rarita-Schwinger operator. In equation (12.32), we have subtracted the contribution from a spin $1/2$ index in $\eta_R/2$ because a Rarita-Schwinger field in dimension twelve decomposes into a Rarita-Schwinger field plus a Dirac field on the boundary $\partial M^{12} = \Sigma^{11}$. Combining (12.31) and (12.32), one obtains the change in the action under a large diffeomorphism:

$$\begin{aligned}
\Delta \mathcal{L}_{\text{eff}} &= 2\pi i \left[\tfrac{1}{2} \eta_D \left(\operatorname{ind} i \slashed{D} \right) + \tfrac{1}{2} N_R \left(\operatorname{ind} (R - S) \right) \right. \\
&\quad \left. - 2 \operatorname{ind} i \slashed{D} - \tfrac{1}{8} N_S \, \sigma \right] \\
&\quad - 2\pi i \int_{M^{12}} \left[\tfrac{N_D}{2} \widehat{A} (M^{12}) + \tfrac{N_R}{2} \left(K (M^{12}) \right) \right. \\
&\quad \left. - 2 \widehat{A} (M^{12}) \tfrac{1}{8} N_S L (M^{12}) \right].
\end{aligned} \qquad (12.33)$$

For the $N = 2$ chiral supergravity theory with $N_D = -2$, $N_R = 2$, $N_S = 1$, the integrands (given by M^{12} curvature) cancel. In twelve-dimensional Euclidean space, charge conjugation does not change the chirality of spinors (see chapter 10 for an explanation); therefore, ind $\not{D}$ and ind $(R - S)$ will always be integers. Hence, these terms can be dropped since $\mod 2\pi i$ they do not contribute to (12.33). Consequently, (12.33) reduces to the final form

$$\Delta \mathcal{L}_{\text{eff}} = 2\pi i \frac{\sigma(M^{12})}{8}. \tag{12.34}$$

How does one infer the absence of global gravitational anomalies when the $N = 2$ supergravity theory is formulated in S^{10}? According to Milnor [6], every exotic 11-sphere Σ^{11} and space $(S^{10} \times S^1)_\pi \simeq (S^{10} \times S^1) + \Sigma^{11}_\pi$ bounds a manifold M^{12} whose signature is divisible by eight, i.e. $\sigma(M^{12}) = 0 \mod 8$. Thus, on the basis of these topological observations,

$$\Delta \mathcal{L}_{\text{eff}} = 0 \mod 2\pi i, \tag{12.35}$$

which is precisely the statement that the $N = 2$, $D = 10$ supergravity theory is free of global gravitational anomalies. Equation (12.35) is referred to as Witten's formula for global anomalies and was initially derived in [4]. We willsee in Subsection 12.2.3 how this formula applies to six-dimensional theories as well.

12.2.2 Global Gravitational Anomalies in the Heterotic String Theory

We will follow the definitions and notations of Section 10.3 of Chapter 10 entitled, the Green-Schwarz Anomaly Cancellation. To briefly recall what we have seen so far in this chapter: global anomalies arise whenever the function space over which we are integrating is not simply connected, thereby preventing the effective action from returning to its original value (mod $2\pi i$), while going through a non-contractible loop in the function space. We will not consider here the heterotic string global world sheet anomaly. For a detailed analysis of this class of anomalies, we refer the interested reader instead to reference [7]. What we will consider though, are the anomalous $E_8 \times E_8$ fields and their overall contribution to the global gravitational anomalies.

What are the heterotic string fields with anomalous terms? They are listed as

(i) the Dirac field, $N_D = 495$;

(ii) the Rarita-Schwinger field, $N_R = 1$.

From the results derived in subsection 12.2.1, a change in the determinant under the large diffeomorphism π in dimension ten is

$$
\begin{aligned}
\Delta \mathcal{L}_{\text{eff}} &= i\pi \left(\tfrac{494}{2}\, \eta_D + \tfrac{1}{2}\, \eta_R \right) \\
&= i\pi \left(494\, \text{ind}_B\, \not{D} + \text{ind}_B\, (R) \right) \\
&\quad - i\pi \int_B \left(493\, \hat{A}\,(R) + K\,(R) \right).
\end{aligned}
\tag{12.36}
$$

The terms $\text{ind}_B\, \not{D}$ and $\text{ind}_B\,(R)$ are respectively the Dirac and Rarita-Schwinger indices on B, a manifold taken to be $B = \partial\,(M \times S^1)_\pi$. Recall that in twelve dimensions, $\text{ind}_B\, \not{D}$ and $\text{ind}_B\,(R)$ are even. We only want the $\Delta\, \mathcal{L}_{\text{det}}$ contribution mod $2\pi i$. The contribution of these terms to (12.36) is trivial and so we may as well drop them. The second point of interest lies in the curvature integral expressed in (12.35): as it stands, it is not a topological invariant since B is endowed with a boundary. In view of this, we rewrite (12.36) as a new formula,

$$
\Delta \mathcal{L}_{\text{det}} = -\frac{i}{(2\pi)^6} \int_B \left(\frac{1}{1536} \left(\text{Tr}\, R^2 \right)^3 + \frac{1}{384} \left(\text{Tr}\, R^2 \right) \left(\text{Tr}\, R^4 \right) \right).
\tag{12.37}
$$

But here too we run into some subtleties. For instance, to produce a topological invariant, we must somehow generalize the terms $\int_B \text{Tr}\, R^6$, $\int_B \left(\text{Tr}\, R^2 \right)^3$ and $\int_B \left(\text{Tr}\, R^2 \right) \cdot \left(\text{Tr}\, R^4 \right)$. There is no known generalization for $\int_B \text{Tr}\, R^6$. But this is hardly an obstacle in the anomalies evaluation since Green and Schwarz have worked mainly with $N_D = 495$, $N_R = 1$, thereby bypassing any need for the term $\text{Tr}\, R^6$.

On the other hand, $\int_B \left(\text{Tr}\, R^2 \right)^3$ and $\int_B \left(\text{Tr}\, R^2 \right) \cdot \left(\text{Tr}\, R^4 \right)$ admit a generalization (that gives rise to topological invariants), provided that, on $(M \times S^1)_\pi$, we solve $\text{Tr}\, R^2 = dH$, $\text{Tr}\, R^4 = dK$. (H and K are defined as in Subsection 12.2.1). Now, the $E_8 \times E_8$ (or equivalently the SO (32)) string theory has a physical field, H, which needs to obey $\text{Tr}\, R^2 = dH$, and thus, as noted by Witten in [5], this very fact puts a strong requirement that $\text{Tr}\, R^2 = dH$ indeed be solved on M. Primarily, the procedure to do so amounts to extending H and K to $(M \times S^1)_\pi$, a rather difficult endeavor. But this can be done, and when successfully completed, formula (12.36) generalizes quite smoothly to a topological invariant, provided that

$$\frac{i}{(2\pi)^6} \int_{(M \times S^1)_\pi} H \left(\frac{1}{1536} \left(\mathrm{Tr}\, R^2 \right)^2 + \frac{1}{384} \mathrm{Tr}\, R^4 \right)$$

is added to it.

Next, consider the purely gravitational term:

$$\begin{aligned} G \;=\; & \frac{i}{(2\pi)^6} \left(\int \frac{1}{1536} P \left(\mathrm{Tr}\, R^2 \right)^2 \right. \\ & \left. + \frac{1}{384} P \, \mathrm{Tr}\, R^4 - \frac{1}{576} \omega_3^L \, \omega_7^L \right). \end{aligned} \tag{12.38}$$

We point out that P denotes the antisymmetric second rank tensor of the supergravity multiplet, whereas ω_3^L and ω_7^L are the Chern-Simons three-and seven-forms. There is a way to extract H, the gauge invariant field strength in terms of P. The relation is

$$H \;=\; dP + \omega_3^L.$$

Under a diffeomorphism transformation evolving from $t = 0$ to $t = 1$, G in (12.38) changes by

$$\begin{aligned} \Delta G \;=\; & \left[\frac{i}{(2\pi)^6} \int_M \left(\frac{1}{1536} \right) P \left(\mathrm{Tr}\, R^2 \right)^2 + \frac{1}{384} P \, \mathrm{Tr}\, R^4 - \frac{1}{576} \omega_3^L \, \omega_7^L \right] \\ =\; & \frac{i}{(2\pi)^6} \int_0^1 dt \int_M \frac{d}{dt} \left[\frac{1}{1536} P \left(\mathrm{Tr}\, R^2 \right)^2 + \frac{1}{384} P \, \mathrm{Tr}\, R^4 - \frac{1}{576} \omega_3^L \, \omega_7^L \right] \\ =\; & \frac{i}{(2\pi)^6} \int_0^1 dt \int_M \left[\frac{1}{1536} (dP) \left(\mathrm{Tr}\, R^2 \right)^2 \right. \\ & \left. + \frac{1}{384} (dP) \, \mathrm{Tr}\, R^4 - \frac{1}{576} d \left(\omega_3^L \, \omega_7^L \right) \right]. \end{aligned} \tag{12.39}$$

Using the formal 12-form expression [4] for the anomaly (e.g. Chapter 10, Section 10.3 and [8]) and taking into account the regulator factor exhibited in [5], we obtain:

$$\begin{aligned} \Delta \mathcal{L} \;=\; & i(2\pi)^6 \int_0^1 dt \int_M \left[\frac{1}{1536} \omega_3^L \left(\mathrm{Tr}\, R^2 \right)^2 \right. \\ & + \left[\frac{1}{1152} \omega_3^L \, \mathrm{Tr}\, R^4 + \frac{1}{576} \mathrm{Tr}\, R^2 \, \omega_3^L \right] \\ =\; & \frac{1}{(2\pi)^6} \int_1^0 dt \int_M \left[\frac{1}{1536} \omega_3^L \left(\mathrm{Tr}\, R^2 \right)^2 \right] \\ & \left. + \frac{1}{384} \omega_3^L \, \mathrm{Tr}\, R^4 + \frac{1}{576} d \left(\omega_3^L \, \omega_7^L \right) \right]. \end{aligned} \tag{12.40}$$

Combining (12.39) and (12.40) yields

$$\Delta \mathcal{L}_{\text{reg}} = \Delta G + \Delta \mathcal{L} = \frac{i}{(2\pi)^6 \cdot 48} \int_{(M \times S^1)_\pi} H$$
$$\cdot \left[\frac{1}{32} \left(\text{Tr} \, R^2 \right)^2 + \frac{1}{8} \text{Tr} \, R^4 \right]; \tag{12.41}$$

where we have used the equivalence relation $H = dP + \omega_3^L$. By combining equations (12.37) with (12.39) and (12.40) one derives Witten's expression for the change in the effective action in terms of topological invariants, that is,

$$\begin{aligned}
\Delta \mathcal{L}_{\text{tot}} &= \Delta \mathcal{L}_{\text{det}} + \Delta G + \Delta \mathcal{L}_{\text{reg}} \\
&= 2\pi i \left[\frac{1}{(2\pi)^6} \int_B \left(\frac{1}{1536} \left(\text{Tr} \, R^2 \right)^3 + \frac{1}{384} \text{Tr} \, R^2 \, \text{Tr} \, R^4 \right) \right. \\
&\quad \left. - \int_{(M \times S^1)_\pi} H \cdot \left(\frac{1}{1536} \left(\text{Tr} \, R^2 \right)^2 + \frac{1}{384} \text{Tr} \, R^4 \right) \right] \\
&= -2i\pi \cdot \frac{1}{192} \left[-3p_1^3 (B) + 4p_1 p_2 (B) \right],
\end{aligned} \tag{12.42}$$

(the p_i are Pontrjagin classes.) When evaluating the anomalies, it is important that the right-hand side of (12.42) be defined only on $(M \times S^1)_\pi$, and moreover, that it be independent of the choice of B.

We are now in position to conclude whether the $O(32)$ and/or $E_8 \times E_8$ superstring theories suffer from global gravitational anomalies when defined in S^{10}. The starting point is the eleven-dimensional spin manifold Σ^{11}, to which we associate the form:

$$\mu(M) = \frac{1}{192} \left[-3p_1^3 + 4p_1 p_2 \right] \mod 1. \tag{12.43}$$

Observe that $(S^{10} \times S^1)_\pi$ is the connected sum of Σ^{11} and $(S^1 \times S^{10})$, which we write as

$$(S^{10} \times S^1)_\pi = \Sigma^{11} \, \natural \left(S^1 \times S^{10} \right).$$

Therefore, the relation follows that

$$\mu \left((S^{10} \times S^1)_\pi \right) = \mu(\Sigma^{11}) + \mu(S^1 \times S^{10}).$$

But $\mu(\Sigma^{11}) = 0$ since any 11-exotic sphere bounds a parallelizable manifold [9, 10, 11]. The Pontrjagin classes p_1 and p_2 then vanish and so $\mu(S^1 \times S^{10})$ is trivial since $S^1 \times S^{10}$ bounds $D^2 \times S^{10}$ which has $p_1^3 = p_1 p_2 = 0$ as well. D^2 is a two-dimensional disc. For the theories in question to be global gravitational anomaly-free under a large change of diffeomorphism π, their effective actions must be invariant under π. This consistency condition is satisfied whenever the formula

$$\mu = 0 \bmod 1 \qquad (12.44)$$

holds.

12.2.3 Global Gravitational Anomalies in Some $D = 6$ Supergravity Theories

We shall analyze here the manifestation of global anomalies generated by disconnected general coordinates transformations. Specifically, we will investigate whether global gravitational anomalies occur in the $N = 2$, $D = 6$ as well as the $N = 4$, $D = 6$ supergravity theories. The impetus to study global anomalies in six dimensions lies in superstring theory, which, in order to make contact with 4-dimensional phenomenology, requires the ten-dimensional theory to break down into product of four-and six-dimensional submanifolds. String theory compactification requires the three-dimensional (complex) submanifold in question to be a Calabi-Yau manifold (see reference [12] for a detailed explanation of this, and additionally, reference [13] for a case study of compactified Calabi-Yau manifolds for the heterotic string theory). On the other hand, supergravity theories are strong candidates for describing low-energy limits of the superstring theory. It is therefore important, for consistency purposes, that supergravity theories be free of global anomalies. As we have discussed throughout this chapter and the preceding ones, there are really two ways to achieve consistency: one is to make sure that the theory we are working with is perturbatively (and chiraly) anomaly-free, and the other complementary option is to cancel any existing global anomalies. Consistency conditions impose such stringent restrictions. This is perhaps why anomalies are seen as *good things* [17]: they restrict the number of possible consistent theories in any given dimension and are hence a powerful tool in the pursuit of uniqueness.

Our approach is as follows. First, we will apply Witten's formula for global anomalies (of Subsection 12.2.1) to the six-dimensional class of theories considered here. The framework in which we will operate draws on the work of Bergshoeff et al. [14], whom, as far as I know, were the first to investigate global gravitational anomalies in six-dimensional supergravity theories. A few comments. The reader may wonder why we are not considering here the $N = 8$, $D = 6$ supergravity theory. The reason for this is simply that it behave in a vector-like fashion with respect to the gravitational interaction [15], thus making the extraction of information on the anomalies side almost impossible. As for the $N = 6$, $D = 6$ theory, when including the field contents, it suffers from perturbative anomalies which cannot be cancelled by the Green-Schwarz mechanism (see Chapter 10, section 10.3 for an exhaustive account of this mechanism.) This very fact therefore rules out the $N = 6$, $D = 6$ supergravity theory.

A. The Case of the $N = 4$, $D = 6$ Supergravity Theory

Our initial focus is the perturbative anomaly. The theory has two basic $N = 4$ chiral multiplets [16] derived from

$$
\begin{array}{ll}
g_{\mu\nu}, \ \Psi^i_\mu, \ B^{+[ij]}_{\mu\nu} & \text{supergravity multiplet} \\
B^{-}_{\mu\nu}, \ \chi, \ \Phi^{[ij]} & \text{tensor multiplet.}
\end{array}
\tag{12.45}
$$

The gravitino in (12.45) is left-handed. The spinor of the tensor multiplet is a right-handed Majorana-Weyl gravitini. The field strength $B^+_{\mu\nu}$ is self-dual.

Perturbative anomalies are characterized by an 8-form polynomial P. We take into account the contributions of a complex left-handed spinor, a complex left-handed Rarita-Schwinger (R-S) spinor and a real self-dual tensor field whose explicit form, taken from reference [4], is

$$
\begin{aligned}
I_{1/2} &= \widehat{A}(R) = \tfrac{1}{5760} \left[\operatorname{Tr} R^4 + \tfrac{5}{4} \left(\operatorname{Tr} R^2 \right)^2 \right]; \\
I_{3/2} &= \widehat{A}(R) \left[\operatorname{Tr}(\cos R - 1) + D - 1 \right] \\
&= \widehat{A}(R) \left(\operatorname{Tr} \cos R - 1 \right) - 2\widehat{A}(R) \\
&= \tfrac{1}{5760} \left[245 \operatorname{Tr} R^4 - \tfrac{215}{4} \left(\operatorname{Tr} R^2 \right)^2 \right]; \\
I_1 &= -\tfrac{L(R)}{8} = \tfrac{1}{5760} \left[28 \operatorname{Tr} R^4 - 10 \left(\operatorname{Tr} R^2 \right)^2 \right].
\end{aligned}
\tag{12.46}
$$

In these quantities, $\widehat{A}(R)$ is defined as before, $L(R)$ is the Hirzebruch polynomial, while D stands for the dimension of M, the space-time manifold of interest. According to the formula above, there are four left-handed Majorana-Weyl gravitini, $4k$-right-handed Majorana-Weyl spinors χ, and k tensor multiplets. The total anomaly polynomial therefore reads

$$
\begin{aligned}
P_{\text{total}} &= 2I_{3/2} - 2k\, I_{1/2} - (k-5)\, I_1 \\
&= \tfrac{(k-21)}{5760} \left[-30 \operatorname{Tr} R^4 + \tfrac{15}{2} \left(\operatorname{Tr} R^2 \right)^2 \right].
\end{aligned}
\tag{12.47}
$$

Townsend has shown [16] that for $k = 21$, the theory is perturbatively anomaly-free. The Hirzebruch signature derived from formula (12.46),

$$\sigma = -\mathrm{ind}\,(R - S) + 23\,\mathrm{ind}\,\not{D}, \qquad (12.48)$$

hold for any closed eight-dimensional spin manifold. Moreover, we take note of the fact that

$$\mathrm{ind}\,\not{D} = \int_M \hat{A}(R);$$

$$\mathrm{ind}\,(R - S) = \int_M \hat{A}(R)\,(\mathrm{Tr}\cos R - 1);$$

$$\sigma = \int_M L(R).$$

Next, the focus is on global anomalies. For convenience, let us recall some basic facts which were the basis for our discussion in Subsections 12.2.1 and 12.2.2. Global gravitational anomalies arise whenever the effective action changes under a large diffeomorphism $(M)_\pi$ that is not continuously connected to the identity. We write this change as

$$\Delta\mathcal{L}_{\mathrm{eff}} = \mathcal{L}_{\mathrm{eff}}\,(g^\star_{\mu\nu}) - \mathcal{L}_{\mathrm{eff}}\,(g_{\mu\nu}),$$

and Witten's formula for global anomalies [5] tells us that a given theory is global anomaly-free if $\Delta\mathcal{L}_{\mathrm{eff}} = 0 \bmod 2\pi i$.

Consider a six-dimensional supergravity theory with N_D left-handed Majorana-Weyl spinors, N_R left-handed gravitini, and N_S self-dual tensor fields. Applying Witten's formula gives

$$\Delta\mathcal{L}_{\mathrm{eff}} = \frac{i\pi}{2}\left[N_D\,\eta_D + N_R\,(\eta_R - \eta_D) - N_S\,\eta_S/2\right]. \qquad (12.49)$$

We have used the Atiyah-Patodi-Singer theorem to find

$$
\begin{aligned}
\tfrac{1}{2}\eta_D &= \mathrm{ind}\,\not{D} - \int_B I_{1/2} \\
\tfrac{1}{2}\eta_R &= \mathrm{ind}\,(R - S) - \mathrm{ind}\,\not{D} - \int_B \left(I_{3/2} + I_{1/2}\right) \\
\eta_S &= \sigma - \int_B L(R).
\end{aligned}
\qquad (12.50)
$$

In the $N = 4$, $D = 6$ supergravity coupled to 21 tensor multiplets, we have $N_D = -4 \times 21$, $N_R = 4$ and $N_S = 5 - 21$. On the basis of this, we combine (12.48) with (12.47) and obtained the result

$$
\begin{aligned}
\Delta\mathcal{L}_{\mathrm{eff}} &= 2i\pi\left[-46\,\mathrm{ind}\,\not{D} + 2\mathrm{ind}\,(R - S) + 2\sigma\right] \\
&\quad - 2\pi i \int_B \left[-42\,I_{1/2} + 2I_{3/2} - 16I_1\right].
\end{aligned}
\qquad (12.51)
$$

Observe that the integrand in (12.51) contains terms which are characteristic of perturbative anomalies, as exhibited in formula (12.49), and therefore vanish. Moreover, $\Delta \mathcal{L}_{\text{eff}}$ is an invariant mod $2\pi i$ because the terms in the bracket in formula (12.51) are multiples of integers. This establishes the absence of global gravitational anomalies in the $N = 4$, $D = 6$ chiral supergravity theory.

B. The $N = 2$, $D = 6$ Supergravity Theory Case

This theory contains the following multiplets:

$$
\begin{array}{ll}
g_{\mu\nu}, \, \Psi_\mu^A, \, B_{\mu\nu}^+ & \text{supergravity} \\
B_{\mu\nu}^-, \, \chi^A, \, \phi & \text{tensor} \\
A_\mu, \, \lambda^A & \text{Yang-Mills} \\
\Psi^a, \, \phi^a & \text{hypermatter.}
\end{array}
\tag{12.52}
$$

The Ψ_μ^A and λ^A are left-handed, while Ψ^a and χ^A are right-handed. According to (12.47), to cancel the leading perturbative gravitational anomaly (which is proportional to $\text{Tr}\, R^4$) requires the consistency condition [14]:

$$
\dim G = n - 29k + 273 = 0.
\tag{12.53}
$$

We then have to worry about the gravitational anomaly:

$$
\frac{k - 9}{128} \left(\text{Tr}\, R^2 \right)^2 .
\tag{12.54}
$$

For $k \neq 9$, one cancels (12.54) by the use of the Green-Schwarz mechanism. There is, however, the pure gauge anomaly to be taken into account. It is given by the term $\text{Tr}\, A^4/4!$ where A is the Yang-Mills curvature corresponding to G.

Hence, to actually cancel the (perturbative) anomalies, either $\text{Tr}\, A^4$ must vanish or be factorized as

$$
\text{Tr}\, A^4 = \sum_1^r \text{Tr}\, F_\mu \, \text{Tr}\, F^\mu,
\tag{12.55}
$$

in order for the Green-Schwarz mechanism to work. The resulting mixed anomaly is

$$
\frac{1}{96} \text{Tr}\, R^2 \, \text{Tr}\, A^2.
\tag{12.56}
$$

The total anomaly polynomial, obtained by combining equations (12.54-55-56), reads

$$P_{\text{total}} = \sum_{i=1}^{r} \beta_{ij} \operatorname{Tr} F_i^2 \operatorname{Tr} F_j^2. \tag{12.57}$$

The term β_{ij} denote an $(r+1) \times (r+1)$ symmetric matrix; $F^{(0)}$ stands for a given Lorentz algebra-valued Riemann curvature 2-form. Under the spell of the Green-Schwarz mechanism, β_{ij} takes on the form:

$$\beta_{ij} = \frac{1}{2} \left(\alpha_i \gamma_j + \alpha_j \gamma_i \right).$$

We cancel the anomaly by adding the counterterm

$$\begin{aligned} \Delta \mathcal{L}'_{\text{eff}} &= \sum_{i=1}^{r} \gamma_i \operatorname{Tr} F_i^2 H \\ &\quad - \tfrac{1}{2} \sum_{i=1}^{r} \alpha_i \gamma_j \omega_{ij}; \end{aligned} \tag{12.58}$$

here, ω_{ij} are the Chern-Simons forms, and $g_{\mu\nu}$ and H denotes the combination of $H_{\mu\nu}^+$ and some $k\, H_{\mu\nu}^-$ tensor fields.

In the $N = 2$ case, the global gravitational anomaly receives two additional contributions, in addition to that of the right hand side of equation (12.51). Setting $N_D = 2$, $N_R = 2$ and $N_S = 1$ gives

$$\begin{aligned} \Delta \mathcal{L}_{\text{det}} &= 2\pi i \left[\operatorname{ind} R + (28k - 274) \operatorname{ind} D + (k-1) \tfrac{\sigma}{8} \right] \\ &\quad - 2\pi i \int_B \sum \beta_{ij} \operatorname{Tr} F_i^2 \operatorname{Tr} F_j^2. \end{aligned} \tag{12.59}$$

Adding the variation of the Green-Schwarz counterterm (12.58), and the Pauli-Villars regulator fields does contribute to (12.59) as follows:

$$\Delta \mathcal{L}'_{\text{eff}} + \Delta \mathcal{L}_{\text{reg}} = -2\pi i \int_{\partial B} Z \sum_{i=1}^{r} \gamma_i \operatorname{Tr} F_i^2. \tag{12.60}$$

The total global gravitational anomaly is then easily derived following the presentation given in the above Subsection 12.2.2. Explicitly,

$$\begin{aligned} \Delta \mathcal{L}_{\text{eff}} &= \Delta \mathcal{L}_{\text{det}} + \Delta \mathcal{L}'_{\text{eff}} + \Delta \mathcal{L}_{\text{reg}} \\ &= 2\pi i \left[(k-1) \tfrac{\sigma}{8} + \mu(B, \partial B) \right] \bmod 2\pi i. \end{aligned} \tag{12.61}$$

Note that μ is a topological invariant whose explicit form is given by

$$\mu(B, \partial B) = \int_B \sum \beta_{ij} \operatorname{Tr} F_i^2 \operatorname{Tr} F_j^2 - \int_{\partial B} Z \sum \gamma_i \operatorname{Tr} F_i^2.$$

Furthermore, Z is the gauge invariant strength, which we write as $Z = dH + \sum_{i=1}^{r} \alpha_i \omega_i$, and

$$dZ = \sum_{i=1}^{r} \alpha_i \, \mathrm{Tr} \, F_i^2.$$

12.3 References

[1] Bott, R. and Tu, L. W.: **Differential Forms in Algebraic Topolgy**, Springer-Verlag 1982 New York.

[2] Witten, E.: *An* SU (2) *Anomaly*, Physics Letters B 117 No.5 (1982) 324-328.

[3] Atiyah, M. F. and Singer, I. M.: Annals of Mathematics 93 (1971) 119.

[4] Alvarez-Gaumé, L. and Witten, E.: *Gravitational Anomalies*, Nuclear Physics B234 (1983) 269-330.

[5] Witten, E.: *Global Gravitational Anomalies*, Commun. Mathematical Physics 100 (1985) 197-229.

[6] Milnor, J. W.: **Differentiable Manifolds Which Are Homotopy Spheres**, Princeton University Press 1959 Princeton, NJ.

[7] Witten, E.: *Global Gravitational Anomalies in String Theory*, in **Proceedings of Symposium in Anomalies, Geometry and Topology**, Eds. Bardeen, W. and White, A., World Scientific 1985 p. 61-99. New Jersey, London, Singapore.

[8] Alvarez-Gaumé, L. and Ginsparg, P.: Nuclear Physics B 243 (1984) 449.

– Annals of Physics 161 (1985) 423.

[9] Baadhio, R. A.: *Global Gravitational Instantons and Their Degree of Symmetry*, Journal Mathematical Physics 33 No. 2 (1992) 721-724.

– *On the Global Gravitational Intantons that Are Homotopy Spheres*, Journal Mathematical Physics 32 No. 10 (1991) 2869-2874.

[10] Baadhio, R. A.: *Knot Theory, Exotic Spheres and Global Gravitational Anomalies*, in **Quantum Topology**, by Kauffman, L. H. and Baadhio,

R. A.; p. 78-90, World Scientific 1993.

[11] Baadhio, R. A. and Kauffman, L. H.: *Link Manifolds and Global Gravitational Anomalies*, Reviews in Mathematical Physics Vol. 5, No. 2 (1993) 331-343.

[12] Green, M.; Schwarz, J. H. and Witten, E.: **Superstring Theory**, Cambridge University Press 1987, New York.

[13] Baadhio, R. A.: *Heterotic Superstring Gauge Residue Trivialization Via Homogeneous CP^4 Topology Change*, Commun. Math. Physics 136 (1991) 251-264.

[14] Bergshoeff, E.; Kephart, T. W.; Salam, A. and Sezgin, E.: *Global Anomalies in Six Dimensions*, Modern Physics Letters A Vol. 1 No. 4 (1986) 267-276.

[15] Tanni, Y.: Phys. Letters B 145 (1984) 197.

[16] Townsend, P. K.: Phys. Letters B 139 (1984) 283.

[17] Gross, D.: *The Heterotic String*, in **Symposium in Anomalies, Geometry and Topology**, Eds. Bardeen, W. A. and White, R. A., World Scientific, 1985 p. 299-313. New Jersey, Singapore, London.

Chapter 13

Mapping Class Groups and Global Anomalies

Throughout this book, we have highlighted the role played by mapping class groups in various physical problems. In this chapter, we will explicitly work out the relationship between global anomalies and mapping class groups in three-dimensional topological quantum field theories, also known as Chern-Simons-Witten theories. Our study will draw on the work of Baadhio from reference [14]. We will begin with a presentation for some three-dimensional homeotopy groups of 3-manifolds which have distinct topologies. The basis for this will be Baadhio's work in reference [20]. Then, in a second phase, we study the occurrence of global gravitational anomalies and the role that mapping class groups have in cancelling them. In doing so, we will follow the presentation exhibited in reference [14]. Our knowledge of mapping class groups of dimensions higher than two is very limited. Nonetheless, drawing on the work in reference [20], we will provide a detailed account of 3-dimensional homeotopy groups. An essential ingredient of the present study is the Smale-Hatcher conjecture.

13.1 The Smale-Hatcher Conjectures

The Smale conjecture and the Smale-Hatcher conjecture are essential ingredients in our investigation of three-dimensional mapping class groups. We thus begin by reviewing them. To state the Smale conjecture for the two-dimensional case, we begin by choosing a properly-imbedded two-dimensional disc in M_g, a three-manifold of genus $g \geq 2$, endowed with at least one boundary component. It is a well known fact [1] that the diffeomorphism group of D^2, when restricted to its boundary component ∂D^2, yields a diffeomorphism f, which is the identity; in other words:

$$\mathrm{Diff}\left(D^2\,\mathrm{rel}\,(\partial D^2)\right) = \{f : D^2 \rightarrow D^2 : f|\partial D^2 = \mathrm{id}\}.$$

Smale conjectured in [2] that the mapping class group of D^2 is trivial:

$$\pi_0\,\mathrm{Diff}^+\left(D^2\,\mathrm{rel}\,(\partial D^2)\right) = 0.$$

He arrived at this conclusion by studying the inclusion of the orthogonal group $O(n)$ into $\mathrm{Diff}(S^n)$, the diffeomorphism group of the n-sphere with $\mathbb{C}^\infty$ topology. The conjecture is affirmatively solved whenever the inclusion is shown to be a homotopy equivalence. In actual life, there is an abundance of forms of the Smale conjecture; the best known example is that of

$$\mathrm{Diff}(S^n) \simeq O(n+1) \times \mathrm{Diff}\left(D^n\,\mathrm{rel}\,(\partial D^n)\right),$$

for any n.

The next in order of priority is a generalization of the Smale conjecture to dimension three. This is referred to as the Smale-Hatcher conjecture. According to this conjecture, for any D^3 with one boundary component, its mapping class group is trivial as well:

$$\pi_0\,\mathrm{Diff}^+\left(D^3\,\mathrm{rel}\,(\partial D^3)\right) = 0.$$

This result follows mostly from Cerf [3], Smale [2], and Hatcher [4].

To complete the present background review, let us point out some useful results by Kirby-Siebenmann and McCullough-Miller [5] which will be of use later:

$$\mathrm{Diff}\left(M_g\,\mathrm{rel}\,(\partial M_g)\right) \overset{\alpha}{\hookrightarrow} \mathrm{homeo}\left(M_g\,\mathrm{rel}\,(\partial M_g)\right),$$

where α is an inclusion on all π_i if $n \leq 3$. Note that under suitable conditions, this relation equates diffeomorphism groups to homeomorphism groups. As such, they shall prove essential in any investigation of Dehn slides in the study of mapping class groups.

13.2   Mapping Class Groups For 3-Handlebodies

This section, which is rather technical, deals with the algebraic topology determination and spresentation of mapping class groups for three-dimensional handlebodies. Specifically, we will be looking at the 3-handlebody with only one boundary component first, then in the later case study that of the 3-handlebody with no boundary component at all. The case with one boundary component yields a trivial mapping class group. This renders any evaluation of global anomalies somewhat difficult for reasons that are explained in Subsection 13.4.1. Three-dimensional handlebodies, in which topological quantum field theories live, offer us an additional challenge: the lack of a one-dimensional time component (in contrast to non-proper 3-manifolds which are analyzed next). It would take us beyond the scope of this chapter to analyze them. The interested reader may want to look at the work in reference [20] for a perspective on this, including quantum gravity in its three-dimensional incarnation.

13.2.1   Homeotopy Groups for 3-Handlebodies With One Boundary Component

We begin with a theorem of Baadhio [20].

Theorem 13.1 (Baadhio) *Consider a three-dimensional handlebody H_g of genus $g \leq 2$. It possesses one boundary component with the identity on the boundary fixed pointwise by* rel ∂H_g. *We then have:*

$$\pi_0 \, \text{Diff}^+ \, (H_g \, \text{rel} \, \partial H_g) \, = \, 0. \tag{13.1}$$

The proof of **Theorem 13.1** will occupy the remainder of this section. We shall rely on involutions within H_g, on the Smale-Hatcher conjecture and on a surgery formula to establish that the mapping class group of H_g with such a topology is indeed trivial.

As a first step, notice that the interior of H_g consists of a two-dimensional Riemann surface, Σ_g. Furthermore, there is a homeomorphism, called h, which takes H_g to itself. This homeomorphism becomes the identity on $\partial_{\text{int}} H_g = \Sigma_g$; by ∂_{int} we mean the interior of H_g. A sketch of H_g together with the isotope h is provided in Figure 13.1.

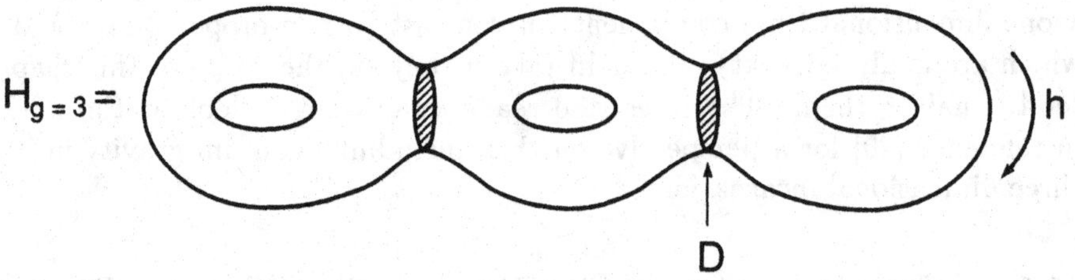

Figure 13.1: Sketched here is a solid 3-handlebody of genus $g = 3$ with one boundary component.

The isotopy h satisfies the following important properties:

$$h_0 = \text{id}$$
$$h|D = \text{id} \text{ ,}$$

where D is a two-dimensional, properly embedded disc in H_g; incidentally, it is depicted in Figure 13.1. We now put forward some useful definitions and notions. By a basic slide homeomorphism within H_g, we shall mean an involution involving h which consists of sliding the disc D in by the trace of h. More generally, any homeomorphism of H_g that uses the disc D and a path c will be referred to as a slide homeomorphism or a slide of the sliding disc D around the sliding path c. When the sliding path c is a well-defined arc however, the slide homeomorphism will be called a Dehn slide. With these descriptions in hand, we can now proceed with the proof of **Theorem 13.1.**

As a first step, we slighty perturb h by a small isotopy to the extent that $h(D)$ involutions intersect transversally. Figure 13.2 provides an illustration of the procedure in question. Another perturbation of $h(D)$ yields circles that are either of the intersection type $\partial D = \partial h(D)$ or not. There are far more circles than we need to care about since we are interested only in identity-preserving involutions, that is, in intersections of the type $D \cap h(D)$. Therefore, a crucial task is to reduce the number of unwanted circles. To this end, we consider the innermost circles of $D \cap h(D)$ which are exhibited in Figure 13.3.

As $h(D)$ sweeps through D and H_g, it does indeed generate a collection of bold black and regular black discs pictured in Figure 13.4. Their unions are precisely the spheres we are seeking to reduce.

A good look at Figure 13.4 tells us that inside the disc there are some balls, B_i. To selectively eliminate the unwanted circles, we use a variant of a theorem of Alexander [6, 7]; this theorem asserts that every properly-embedded two-dimensional disc (or sphere for that matter) in H_g is indeed the boundary of a three-dimensional ball [1]. Pushing the (bold) discs across the balls in Figure 13.4 allows one to selectively eliminate the corresponding

---

[1]There does not seem to exist a satisfactory reference for Alexander's theorem for the smooth case. It has been curiously neglected in some of the standard books on 3-manifolds. There is however, a rather simple and beautiful proof by Hatcher in unpublished notes. I thank him for bringing this to my attention.

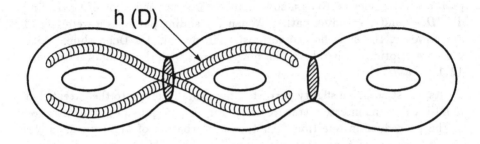

Figure 13.2: Pointwise boundary-fixed preserving homeomorphism involutions on H_g generated by small perturbations of the disc D.

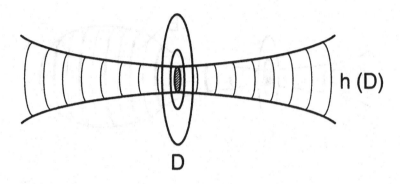

Figure 13.3: Cross view of arcs and circles generated by the intersections $D \cap h(D)$ from $h(D)$ involutions.

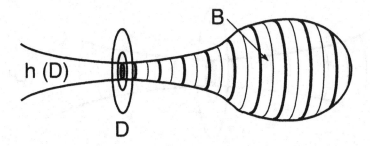

Figure 13.4: Balls and discs generated by small perturbations of $h(D)$.

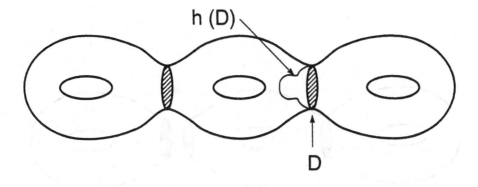

Figure 13.5: Pushing the isotope $h(D)$ to agree with D.

circles made up of $D \cap h(D)$. We repeat the operation as necessary to eliminate the remaining circles.

The second step consists of pushing $h(D)$ to agree with D, once it is established that $D \cap h(D) = \partial D = \partial h(D)$. A sketch of this procedure is provided in Figure 13.5. This amounts to a net decrease in the amount of slide homeomorphism involutions. The actual framework that allows us to achieve this result owes much to the Lemma 4.4 by Suzuki in reference [8].

By restricting $h(D)$ to the identity, that is to $h|D = \text{id}$, we can perform a series of surgeries on H_g. We begin by cutting the 3-handlebody along D, as pictured in Figure 13.6.

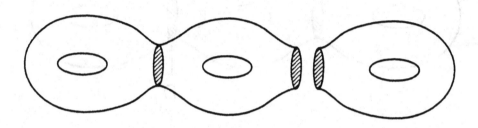

Figure 13.6: Initial surgery on H_g along D with induction on the genus g.

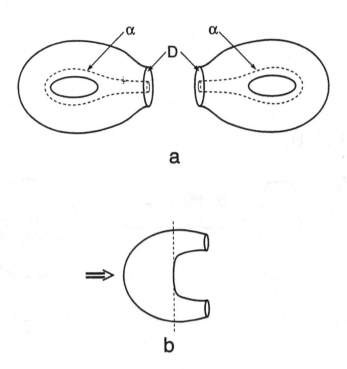

Figure 13.7: Repeated surgery on H_g with induction on g.

This surgery procedure has an induction on the value of g: one keeps cutting H_g along D_i (Figure 13.7).

This continuous use of surgery as depicted in Figure 13.7 reduces the H_g genus size to $g = 3$ to $g = 2$ to $g = 1$, and finally to $g = 0$. The final result is that the handlebody H_g is reduced to D^3, as shown in Figure 13.8.

It is then just a matter of procedure to apply the Smale-Hatcher conjecture to show that

$$\pi_0 \, \text{Diff}^+ \left(H_g \text{rel} \, \partial H_g \right) \equiv \pi_0 \, \text{Diff}^+ \left(D^3 \text{rel} \, \partial D^3 \right) = 0. \qquad (13.2)$$

This completes the proof of **Theorem 13.1**.

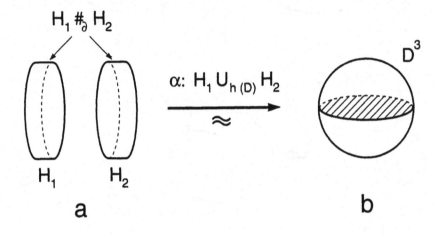

Figure 13.8: Repeated surgeries on H_g have reduced it to a 3-disc, D^3. The Smale-Hatcher conjecture is then applied to determine the proper value of its homeotopy group.

13.2.2 Homeotopy Groups of 3-Handlebodies With No Boundary Component

The principal result is contained in the following theorem of Baadhio [20].

Theorem 13.2 (Baadhio)

$$\pi_0 \, \mathrm{Diff}^+ \, (H_g) \, = \, \Gamma_\star \, \subset \, \pi_0 \, \mathrm{Diff}^+ \, (\Sigma_g, \star). \tag{13.3}$$

$\Gamma_\star$ is the smallest invariant subgroup of the mapping class group, $\pi_0 \, \mathrm{Diff}^+ \, (\Sigma_g, \star)$; incidentally, it is a twist group of finite index, generated by canonical twist automorphisms. As for $\star$, it denotes a basepoint in H_g while the group of $\star$-isotopy classes of automorphisms that keeps $\star$ fixed is referred to as Diff $(\Sigma_g, \star)$.

The first order of business in proving **Theorem 13.2** is to find the canonical twist automorphisms mentioned earlier. This means looking at the automorphism group of π_1, where π_1 is the fundamental group of H_g, a free group of rank n with a $\star$ base with coordinates $x_1, \cdots, x_n$. The automorphisms of the fundamental group give important information about mapping class groups of 3-manifolds including the case of handlebodies. However, finding a presentation for the mapping class group of a 3-handlebody would require additional information about how the Dehn twists around the discs work. The reason lies in the fact that the twists induce the identity automorphism on the fundamental group, and therefore they cannot be detected from the automorphisms of the fundamental group.

The automorphism group of π_1 is generated by:

$$\begin{aligned}
A\,(x_1) &= x_2; \\
A\,(x_2) &= x_1; \\
A\,(x_k) &= x_k; \\
B\,(x_k) &= x_{k+1}; \\
C\,(x_1) &= x_1^{-1}; \\
C\,(x_k) &= x_k; \\
D\,(x_1) &= x_1 x_2; \\
D\,(x_k) &= x_k;
\end{aligned} \tag{13.4}$$

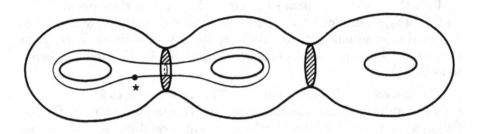

Figure 13.9: Basepoint $\star$-involutions in H_g that are mapping class preserving.

where $k = 1, \cdots, n$ and $x_{n+1} = x_1$. Nielsen's original work on automorphisms of surfaces [9] provides us with the right framework to determine the canonical generators of $\pi_0 \operatorname{Diff}^+ (\Sigma_g, \star) = \Gamma_\star$ on the basis of the A, B, C and D automorphisms in equation (13.4). According to these, when $g \geq 3$, $\Gamma_\star$ has six canonical generators; in contrast, when $g = 2$, there are exactly five canonical generators. In considering the case of interest to us (i.e. $g = 3$), the action on $\pi_1 (H_g)$ of four of the six generators is $A = B, C$ and D. The remaining two generators are defined by canonical twist automorphisms. These twists will prove essential in the Proof of **Theorem 13.2**. It is with this in mind that we aim to determine the twist automorphisms that generate the twist group $\Gamma_\star$.

Firstly, let us clarify what is meant by a twist $\tau (D)$. It is essentially a homeomorphism satisfying the relation:

$$\tau (D) \left(\mathrm{re}^{2\pi i \theta}, t \right) = \left(\mathrm{re}^{2\pi i (\theta + t)} \right).$$

When restricted to ∂H_g, $\tau (D)$ is none other than a Dehn twist in ∂D. The isotopy class $\tau (D)$ in the full mapping class group $\pi_0 \operatorname{Diff}^+ (\Sigma_g, \star)$ can be shown to depend only on the ambient isotopy class of D in H_g. It is precisely the collection of all isotopy classes of twist automorphisms that generates the twist group $\Gamma_\star$, a normal subgroup of the homeotopy group of H_g's interior, Σ_g, with basepoint $\star$. In order to fully understand $\Gamma_\star$, we need to address its finiteness. A previous work on the subject by McCullough (on virtually geometrically finite mapping class groups of 3-manifolds) in reference [10] is of interest. The trick is to make use of the compactness and orientability of H_g, and hence to show in the process that $\Gamma_\star$ is contained in the kernel of the topological isomorphism

$$\pi_0 \operatorname{Diff}^+ (\Sigma_g, \star) \rightarrow \operatorname{Out} (\pi_1 (H_g)), \tag{13.5}$$

or, in the optimal case, is equivalent to that kernel.

What are the conditions for the kernel of the homomorphism in equation (13.5) to be finitely generated? The answer to this obviously depends on whether $\Gamma_\star$ itself is finitely generated. Thus, the whole issue of finiteness amounts to actually seeing whether $\Gamma_\star$ is a geometrically well-defined subgroup (of the mapping class group), and of finite index. Below, we provide some arguments that prove the finiteness of $\Gamma_\star$.

Looking at the 3-handlebody H_g (Figure 13.9), there is a fibration

$$F \ \to E \xrightarrow{\alpha} B \Rightarrow \pi_n(F) \to \pi_n(E) \to$$
$$\pi_n(B) \to \pi_{n-1}(F) \to \pi_{n-1}(E) \to \cdots \qquad (13.6)$$

The exact sequence of homotopy groups in (13.6) is a consequence of the fibration and restriction ρ. According to formula (13.6), we have the sequence

$$\text{Diff}^+ (H_g \operatorname{rel} \partial H_g) \ \to \text{Diff}^+ (H_g) \xrightarrow{\rho} \text{Diff}^+ (\partial H_g); \qquad (13.7)$$

where $\text{Diff}^+ (H_g \operatorname{rel} \partial H_g) = \rho^{-1}(\text{id})$. Mapping class group wise, (13.7) reads:

$$\pi_1 \text{Diff}^+ (\partial H_g) \ \to \pi_0 \text{Diff}^+ (H_g \operatorname{rel} \partial H_g) \to$$
$$\pi_0 \text{Diff}^+ (H_g) \quad \to \pi_0 \text{Diff}^+ (\partial_{\text{int}} H_g = \Sigma_g). \qquad (13.8)$$

As a result of equations (13.6-7) we have the exact sequence

$$0 \to \ker(\alpha) \to \pi_0 \text{Diff}^+ (H_g) \xrightarrow{\alpha} [\text{Aut}(\pi_1(H_g)) = \text{Out}(\text{Aut } H_g)]. \quad (13.9)$$

From (13.9), we learn that $\ker \alpha$ is given by Dehn twists and is not finitely generated. This is in contrast to $\pi_0 \text{Diff}^+ (H_g)$. We regard $\text{Aut}(\pi_1(H_g))$ as acting on the left of $\pi_1(H_g)$ by the action of an injective cyclic group. The term α maps into the inner automorphism, $\text{Aut}(\pi_1(H_g))$; the outer automorphism is a (finitely) generated infinite group. The exact sequence (13.9) yields the final result:

$$\underbrace{\pi_0 \text{Diff}^+ (H_g \operatorname{rel} \partial H_g)}_{=0} \to \underbrace{\pi_0 \text{Diff}^+ (H_g)}_{\neq 0}$$
$$\xrightarrow{\alpha_*} \pi_0 \text{Diff}^+ (\partial_{\text{int}} H_g = \Sigma_g). \qquad (13.10)$$

The first term in (13.10) is trivial by virtue of the theorem of Baadhio, namely **Theorem 13.1**; consequently, the remaining part of **Theorem 13.2** will now involve ρ_*, or more precisely, its kernel. It is a straightforward task to show that ρ_* is injective in (13.9). Thus, as a direct consequence, $\ker \alpha_* = \{0\}$. This, it turns out, implies that the mapping class group $\pi_0 \text{Diff}^+ (H_g)$ is isomorphic to the twist group $\Gamma_* \subset \pi_0 \text{Diff}^+ (\Sigma_g, \star)$.

Our proof of **Theorem 13.2** parallels, and somewhat strengthens, Waldhausen's work on sufficiently large 3-manifolds with incompressible boundaries (nowadays called Haken 3-manifolds) [11]. Waldhausen was among the first to point out an equivalence relation between homeomorphism and homotopy equivalence. An ongoing problem in geometric topology concerns the classification of such homotopy equivalences, particularly for complicated 3-manifolds such as non-aspherical manifolds and manifolds which are non-trivial connected sums [12, 13].

13.2.3   A Presentation For Γ_* In Terms of Dehn Twists

When evaluating (13.10), we need to pay attention to homotopic basepoints in the handlebody, as they easily factor out $\operatorname{Aut} \pi_1(H_g)/\operatorname{Out}\operatorname{Aut}(H_g)$. On the other hand, homotopic basepoints are generators of $\pi_1(H_g)$ and Γ_*. Determining which ones are relevant is the goal of this subsection. The starting point of the discussion is Figure 13.10.

Requiring diffeomorphism extensions of certain Dehn twists in Figure 13.10 is a good method for finding the relevant generators of Γ_*. For instance, the Dehn twist α_1 doesn't extend to $\operatorname{Diff}(H_g)$ since it does not bound a disc. This is not true of α_2 which extends along the disc D_2, and thus to $\operatorname{Diff}(H_g)$. Hence, requiring the infinitely many Dehn twists to bound a disc or equivalently, to extend to $\operatorname{Diff}(H_g)$ is a highly non-trivial approach to separate the *good* Dehn twists from the trivial ones. There are, however additional constraints that need to be satisfied. They have to do with diffeomorphisms of the 3-handlebody that do not induce the identity on $\pi_1(H_g)$. The Dehn twists along the discs D_i induce the identity on $\pi_1(H_g) \to \pi_1(H_g)$, because one can always pick a diffeomorphism $f : H_g \to H_g$ that induces an isomorphism on $\pi_1(H_g) \to \pi_1(H_g)$. In order for us to present the relevant Dehn twists as generators of Γ_*, we need to decide which Dehn twists are the relevant ones, that is, we need to determine the ones that are not identity-inducing in $\pi_1(H_g)$, but are nonetheless homotopic to the remaining Dehn twists.

The procedural approach is outlined as follows. As depicted in Figure 13.11, we apply a 180° rotational twist along the shaded area of H_g. Our goal here is to generate diffeomorphisms of the 3-handlebody that are not identity inducing. A good look at Figure 13.11 tells us that the *half* Dehn twist along $\gamma = \tau_\lambda$ is such a diffeomorphism.

When endowed with the opposite orientation, $\lambda = \tau_\gamma(\lambda)$ is equivalent to

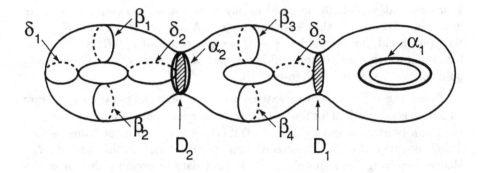

Figure 13.10: Dehn twist involutions in H_g

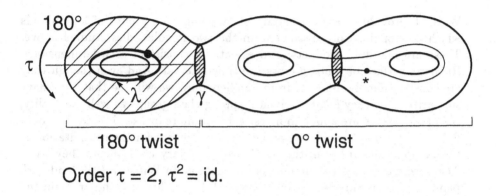

Figure 13.11: Selection rule by twist automorphisms in order to extract Dehn twists that are generators of $\Gamma_\star = \pi_0 \mathrm{Diff}^+ (H_g)$.

λ^{-1} in $\pi_1(H_g)$. These are examples of diffeomorphisms that do not induce the identity on $\pi_1(H_g)$, but are nonetheless homotopic to the other Dehn twists.

13.3   Homeotopy Groups of Non-Proper 3-Manifolds

We shall provide here a case study of mapping class groups of 3-manifolds which are topologically distinct from the two handlebodies studied above. These are manifold products, and essentially consist of a two-dimensional Riemann surface of genus g cross the time interval. And even though they are constitutionally different from handlebodies, we will see that they have similarities, for they exhibit trivial mapping class groups as well. In reality, the manifolds we are about to investigate are more or less similar. The only difference (perhaps trivial overall but very important for the computation of homeotopy groups) lies in the number of boundary components they have. There were the very first manifolds that entered in Witten's 1989 landmark paper on Chern-Simons-Witten theories [18]. To this day, they remain the most widely studied 3-manifolds for Chern-Simons-Witten theories (including quantum gravity).

13.3.1   Mapping Class Groups for $\Sigma_g \times [0, 1]$ With Two Boundary Components

According to the following theorem of Baadhio in reference [20], a given 3-manifold, $\Sigma_g \times [0, 1]$ $(g \geq 2)$ with two boundary components has a non-trivial mapping class group, namely:

Theorem 13.3 (Baadhio)

$$\pi_0 \operatorname{Diff}^+ \left(\Sigma_g \times [0, 1] \operatorname{rel} \Sigma_g \times \{0, 1\} \right) =$$
$$\pi_0 \operatorname{Diff}^+ \left(\Sigma_g \times \mathbb{Z}_2 \right) \approx \pi_1 \operatorname{Diff}^+ \left(\Sigma_g \right). \tag{13.11}$$

In Figure 13.12 we provide a sketch of that 3-manifold.

To prove **Theorem 13.3**, we start by picking an element of $\pi_1 \operatorname{Diff}^+ (\Sigma_g)$. This element is represented by a loop f_t of diffeomorphisms (with $f_0 = f_1 =$

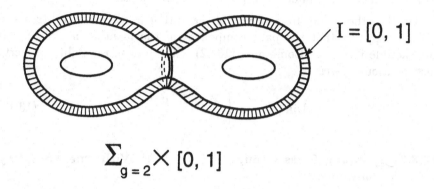

Figure 13.12: Thickened copy of $\Sigma_g \times [0,1]$, or equivalently, $\Sigma_g \times [0,1]$ with two boundary components.

identity on Σ_g); it is defined by $H_g(x,t) = f_t(x)$, which in turn comes from the map

$$\Psi : \pi_1 \left(\text{Diff } \Sigma_g \right) \rightarrow \pi_0 \left(\text{Diff}^+ \left(\Sigma_g \times [0,1], \text{rel } \Sigma_g \times \{0,1\} \right) \right).$$

Thus, we obtain a single diffeomorphism F of $\Sigma_g \times [0,1]$:

$$F \in \text{Diff} \left(\Sigma_g \times [0,1] \, \text{rel } \partial^2 \right) \rightarrow F(x,t) = (f_t(x), t);$$

as shown in Figure 13.12.

An essential ingredient in the proof is the realization that any path component of $\text{Diff}^+ \left(\Sigma_g \times [0,1] \, \text{rel } \Sigma_g \times \{0,1\} \right)$ contains level-preserving elements, that is, elements in the image of Ψ, so Ψ is surjective. On the basis of these observations, we have the relation

$$\pi_0 \operatorname{Diff}^+ \left(\Sigma_g \times [0,1] \operatorname{rel} \partial^2 \right) \xrightarrow{\approx} \pi_1 \operatorname{Diff} (\Sigma_g); \qquad (13.12)$$

which, according to Waldhausen [11], is an isomorphism.

On the other hand, from results of the author in reference [14] (see also [15]), it appears that the path component of the identity in $\operatorname{Diff} (\Sigma_g)$ is contractible if $g > 1$. Combining (13.12) with the contractibility property just mentioned gives:

$$\pi_1 \operatorname{Diff} (\Sigma_g) = \begin{cases} \mathbb{Z} \times \mathbb{Z} & \text{if } g = 1 \\ 0 & \text{if } g > 1 \end{cases}. \qquad (13.13)$$

13.3.2 Mapping Class Groups of $\Sigma_g \times [0,1]$ With One Boundary Component

The statement here is that the mapping class group of a 3-manifold $\Sigma_g \times [0,1]$, endowed with one boundary component, is trivial. This statement is best reflected in **Theorem 13.4** (see [20]):

Theorem 13.4 (Baadhio)

$$\pi_0 \operatorname{Diff}^+ \left(\Sigma_g \times [0,1] \operatorname{rel} \Sigma_g \times [0,\infty) \right) \approx$$
$$\pi_1 \operatorname{Diff}^+ \left(\Sigma_g \times \mathbb{Z}_2 \right); \qquad (13.14)$$

where $\mathbb{Z}_2$ denotes the switching of the two ends of $I = [0,1]$.

Before embarking on the proof of this fourth theorem of Baadhio, let us pause to clarify a topological subtlety: it is slightly incorrect for us to state that $\Sigma_g \times [0,1]$ has only one boundary component. In real life, $\Sigma_g \times [0,1]$ always has two boundary components: $\Sigma_g \times \{0\}$ and $\Sigma_g \times \{1\}$. Hence, when we say of $\Sigma_g \times [0,1]$ that it has only one boundary component, we mean that it is diffeomorphic to $\Sigma_g \times [0,\infty)$.

This established, the proof of (13.14) is rather straightforward and draws its base from Cerf's work on diffeomorphism extensions [3]. In particular, it

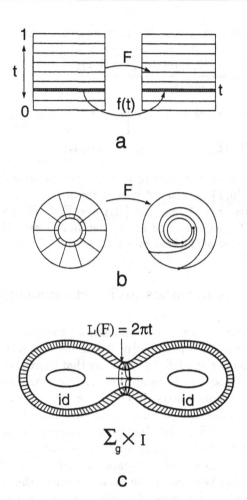

Figure 13.13: Sketched in (a): a path component, $f_t \in \text{Diff}(\Sigma_g)$ which contains level-preserving elements; in (b): a loop within $\text{Diff}(\Sigma_g)$ whose homotopic involution (i.e. a $2\pi t = f_t$ rotation) yields a single isotopic morphism F; in (c): a representation of the involution of F within the annulus A of the 3-manifold, $\Sigma_g \times [0,1]$.

holds true if, and only if, $g > 1$. We have an exact sequence:

$$
\underbrace{\pi_0 \operatorname{Diff}^+ \left(\Sigma_g \times [0,1] \operatorname{rel} \partial \left(\Sigma_g \times [0,1) \right) \right)}_{=0 \text{ if } g>1}
$$

$$
\to \pi_0 \operatorname{Diff}^+ \left(\Sigma_g \times [0,1] \right) \xrightarrow{\alpha} \left[\pi_0 \operatorname{Diff}^+ \left(\Sigma_g \times \underbrace{\{0,1\}}_{=\partial I} \right) \right] \qquad (13.15)
$$

$$
= \left(\pi_0 \operatorname{Diff}^+ (\Sigma_g) \right) \times \left(\pi_0 \operatorname{Diff}^+ (\Sigma_g) \times \mathbb{Z}_2 \right).
$$

Note that ρ is injective and acts to restrict the mapping class group to the boundary. $\Sigma_g \times \{0,1\}$ consists of two copies of Σ_g. The image of ρ is built out of the elements $(x, x) \in \pi_0 \operatorname{Diff}^+ (\Sigma_g) \times \pi_0 \operatorname{Diff}^+ (\Sigma_g) \times \mathbb{Z}_2$. Isotopic diffeomorphisms of Σ_g, as described by Birman in reference [1] are then chosen to complete the proof of **Theorem 13.4**.

13.4   Global Anomalies and 3-Homeotopy Groups

The purpose of this section is to demonstrate the absence of global gravitational anomalies for a small class of 3-manifolds in which Chern-Simons-Witten (CSW) theories are defined. The fact that these theories are pathology-free relies on an intricate relationship between mapping class groups and large diffeomorphism transformations. In effect, we will show that the absence of global anomalies amounts to proving that under large diffeormorphism transformations, the variation of the effective action is continuously connected to the identity, that is, there is no gap or occurrence of disconnected general coordinate transformations. If such gaps existed, one would not expect the variation of the effective action to be invariant under the group of diffeomorphisms that cannot be smoothly deformed to the identity. Rigid deformations of this type are almost exclusively classified (or detected) by homeotopy groups, and this fact alone explains the choice of finding large variations of the effective action which are mapping class group invariant. We will write $\mathcal{L}_{\mathrm{CSW}}$ for the Chern-Simons-Witten effective action. The mapping class group is defined, as above in Section 13.2, with the same notations. For additional references on global anomalies and mapping class groups, we suggest to the interested reader references [16, 17], which are essentially extensions of the work in reference [14].

Consider M^n, a manifold of dimension $n = 3$. In previous sections, when $n = 2 + 1$, we referred to M^n as a non-proper manifold (a product manifold, e.g. a Riemann surface cross $\mathbb{R}^1$), in contrast to $n = 3$ (handlebody case). To this manifold is associated a compact gauge group, G, and a principal bundle, E. There is a connection on E (e.g. a Lie algebra valued one form, $\mathcal{G}$), which we write as A_i^μ, where i is the tangent index to M^n, and μ is defined over a basis of the Lie algebra $\mathcal{G}$. Infinitesimal gauge transformations are of the form:

$$A_i \rightarrow A_i - \nabla_i \epsilon;$$

and ϵ is the generator of G. Another useful ingredient is the covariant derivative, ∇_i

$$\nabla_i \epsilon = \partial_i \epsilon = \partial_i \epsilon + [A_i, \epsilon],$$

which gives rise to an equally important form, the curvature 2-form:

$$F_{ij} = [\nabla_i, \nabla_j] = \partial_i \nabla_j - \partial_j \nabla_i + [\nabla_i, \nabla_j].$$

The Chern-Simons-Witten effective action must be defined as a topological invariant. In so doing, we learn to avoid the standard Yang-Mills Lagrangian,

$$\mathcal{L}_{YM} = \int_{M^n} \sqrt{g}\, g^{ik}\, g^{il}\, \mathrm{Tr}\, (F_{ij}\, F_{kl})$$

since it is metric dependent. A good compromise is to pick the integral of the Chern-Simons form,

$$\begin{aligned} \mathcal{L}_{\mathrm{CSW}} &= \tfrac{k}{4\pi} \int_{M^n} \mathrm{Tr}\, \left(A \wedge dA + \tfrac{2}{3} A \wedge A \wedge A\right) \\ &= \tfrac{k}{8\pi} \int \epsilon^{ijk} \mathrm{Tr}\, \left((A_i\,(\partial_j A_k - \partial_k A_j) + \tfrac{2}{3} A_i\,[A_j, A_k]\right). \end{aligned} \tag{13.16}$$

The formula,

$$\eta = d \cdot \eta_{\mathrm{grav}}$$

in reference [18], determines the purely gravitational operator, η_{grav}. Moreover, it is used to write down the phase factor

$$\Lambda = \exp\left(\frac{id\pi}{2} \cdot \eta_{\mathrm{grav}}\right). \tag{13.17}$$

As it stands, equation (13.17) needs to be regularized if general covariance is to apply; furthermore, we point out its metric dependence since two regularizations differs by a local counterterm. This in turn spells out the need to find a counterterm which is also metric dependent. In reference [18], it is shown that a multiple of the gravitational Chern-Simons term is that ideal counterterm:

$$I_g = \frac{1}{4\pi} \int_{M^n} \text{Tr} \left(\omega \wedge d\omega + \frac{2}{3} \omega \wedge \omega \wedge \omega \right) \qquad (13.18)$$

where the subscript g denotes the background metric dependence, and ω stands for the Levi-Civita connection on M^n. Essentially, the gravitational anomaly counterterm is obtained by indirect derivation, often proceeding via conformal field theory.

On the other hand, there are, so far, only two possible forms for the CSW counterterm: it can be written in terms of a spin connection or an affineconnection [18]. In the first form, the anomalous transformations are the SO $(2, 1)$ gauge transformations (i.e. transformations of the local Lorentz type), while, in the second form, they are diffeomorphisms. The relation between these two forms is very poorly understood [14, 16, 17, 18]. There is no doubt that any progress on this front will provide us with a better undertanding of the occurrence and manifestation of global anomalies in topological quantum field theories.

Let us now review the nature and manifestations of global gravitational anomalies in Chern-Simons-Witten theories . The first symptom is encoded in the η-invariant. Path integral computations carried out by Witten [18] reveals that, to the lowest order of perturbation, the action for a given CSW theory requires a counterterm proportional to the gravitational CSW action. This is precisely the term we encounter in formula (13.18). Since the gravitational CSW action (13.18) is not invariant under large diffeomorphism transformations, this term represents, or otherwise exhibits, a global gravitational anomaly. The second symptom has to do with the the Wilson-line anomaly. This anomaly appears as a framing dependence [18] because the choice of the framing is mapping class group invariant as well. Finally, there is third manifestation which was the subject of a brief presentation in Chapter 8: the canonical quantization of CSW theories makes succinct the relationship between global anomalies and mapping class groups. Roughly speaking, the quantization exhibits an anomaly that shows up as a dependence of the wave function on the moduli space of (non-proper) three-manifolds. The canonical quantization discussed in Chapter 8 essentially gives a projectively flat –but not flat by nature– bundle over the moduli space with a curvature given by the anomaly. The mapping class group $\pi_0 \, \text{Diff}^+ \, (M^n)$ is then known to have a non-trivial action on this bundle. There is an additional anomaly, namely, the Virasoro global gauge anomaly, in its three-dimensional incarnation. This anomaly shows up when one attempts to centrally extend the mapping class group from dimension two to three using the central charge. This peculiar class of global pathologies was initially analyzed in [19] and, as far as we know, would benefit from additional extensive investigations. We will not discuss them here, referring the interested reader instead to [19].

It should be pointed out that the instances of global anomalies detected by the use of mapping class groups dissscussed here are far from representing the whole story. Several approaches aimed at providing a general framework to detect global anomalies in quantum field theories are in the works. One approach, for instance, targets classical invariants of low-dimensional topology and studies their extension to higher dimensions. In higher dimensions, these knot invariants give rise to exotic spheres. As such, the procedure constitutes a powerful test to detect global anomalies simultanously in low dimensions (e.g. three) and in higher ones (e.g. six, seven, eight, ten, and perhaps beyond eleven). This approach was initiated in [17] by Baadhio and Kauffman, with an additional survey article by Baadhio in [16]. What is still lacking though on the topological side is the equivalent of Witten's formula for global anomalies, which would universally inform us of the presence of global anomalies. Research activities are still ongoing though, and hopefully we will find it.

Another approach to the detection and cancellation of global anomalies is that of torsion in mapping class groups. The initial observation that global gravitational anomalies could be understood in terms of torsion is due to Freed [21]. His observation was later applied to the study of the string world sheet global anomalies in [22]. In the mapping class group context, the issue is rendered somewhat manageable by the stability (resp. instability) of the the torsion classes in the homeotopy groups. Torsion stability of mapping class groups could be expressed in terms of the virtual cohomology dimension and the stability range is given by the Harer's stability theorem. Both notions were discussed in Chapter 4 (Section 4.6). An additional tool is perhaps the Farrell-Tate cohomology. This cohomology theory is excellent at detecting some periodicity of the mapping class group. Thus, if somehow combined with Harer's stability theorem, there is no doubt that they will provide us with a powerful tool in the analysis of global gravitational anomalies. One can interpret the unstable torsion class of the homeotopy group as being precisely the group of all disconnected general coordinate transformations, although the correct geometric topology picture of this approach remains to be worked out. In the unstable range, the Farrell-Tate cohomology may act as the marker for the anomalies, with respect to diffeomorphism transformations that are continuously connected to the identity. This may provide us with an important quantization tool and a way to easily cancel the anomalies. Details regarding this method will be reported in [23].

13.4.1   Global Gravitational Anomaly-Free Chern-Simons-Witten Theories

The purpose of this subsection is to establish the absence of global anomalies in Chern-Simons-Witten also known as topological quantum field theories. We will follow the presentation of Baadhio in [14]. Let $\pi_0 \, \text{Diff}^+ \, (M^n)$ denote (non-zero) homeotopy group of a 3-manifold. We ask that (M^n) be either a three-manifold of the type $\Sigma_g \times [0,1]$ or a handlebody. By Σ_g, we mean a two-dimensional Riemannian surface of genus $g \geq 2$, while $[0,1]$ labels the associated one-dimensional time component. In order to demonstrate the absence of global anomalies while varying the effective action $\mathcal{L}_{CSW}$ under large diffeomorphism transformations, we need to show that the resulting changes in the effective action are continuously connected to the identity, that is, the collection of values taken by $\mathcal{L}_{CSW}$ while sweeping through the large diffeormorphisms cannot smoothly be deformed to the identity. When such collections are susceptible to non-smooth deformation with respect to the identity, then the theory suffers from disconnected general coordinate transformations, known simply as global gravitational anomalies.

   The only known diffeormorphism group that cannot be smoothly deformed to the identity (i.e. does not admit general disconnected diffeomorphic transformations) is the mapping class group, or homeotopy group. This is why it is important to establish $\mathcal{L}_{CSW}$ invariance under the mapping class group under those transformations. Not only does the invariance prevent global anomalies from arising, but further, it acts to preserve the consistency and uniqueness of the theory considered. The presence of global anomalies in this case would almost certainly violate the theory's uniqueness. Once again, it is easy to draw the conclusion reached a decade ago by Witten [24] that global anomalies act to severely limit the number of possible consistent theories in any dimension.

A question of interest we are faced with at this point has to do with **Theorem 13.1** and **Theorem 13.4** which exhibit trivial homeotopy groups. What are the mechanisms in place to check the occurrence and manifestations of global gravitational anomalies when the mapping class group is zero? Currently, there is no known answer to this question, but there is little doubt that if answers are found, they will considerably strengthen our understanding.

From the group of orientation preserving diffeormorphisms, $\text{Diff}^+(M^n)$ we extract an element, λ, and require that it obey the relation:

$$\mathcal{L}_{CSW}(\lambda) = \mathcal{L}_{CSW} = \mathcal{L}_{CSW}(\text{id}), \tag{13.19}$$

for all λ; id is the identity. We define h_t, a one-parameter family of λ, by

$$h_t = \lambda_t \ 0 \le t \le 1,$$

to keep track of λ's variation within $\text{Diff}^+(M^n)$. Incidentally, with values between zero and one, we note that

$$h_t \in \pi_0 \text{Diff}^+(M^n) \subset \text{Diff}^+(M^n).$$

The large diffeomorphism transformation

$$\phi \times \text{id} : M^n \to M^n, \tag{13.20}$$

$\left(\phi \in \text{Diff}^+(M^n)\right)$ induces the map

$$M^n \xrightarrow{\phi \times \text{id} \sim \phi^*(A_i)} M^n \xrightarrow{\omega} \mathcal{G}. \tag{13.21}$$

Recall that A_i is the 1-form connection on M^n, and ω its gravitational counterterm; $\mathcal{G}$ is the Lie algebra in which the connection takes its values. The term $\phi^*(A_i)$, is by definition, equivalent to $\phi^*(\omega)$ and is obtained by pulling back along the formula (13.21).

Next, we replace A_i and ω by the variation $\phi^\star (A_i)$ (or similarly, by $\phi^\star (\omega)$). As a consequence, the effective action in formula (13.16) changes by

$$
\begin{aligned}
\mathcal{L} \left(\phi^\star (A_i) \right) &= \tfrac{1}{4\pi} \int_{M^n} \mathrm{Tr} \left[\phi^\star (A_i) \wedge d\phi^\star (A_i) \right. \\
&\quad + \tfrac{2}{3} \phi^\star (A_i) \wedge \phi^\star (A_i) \wedge \phi^\star (A_i) \Big] \\
&= \tfrac{k}{8\pi} \int_{M^n} \epsilon^{ijk} \, \mathrm{Tr} \left[\phi^\star (A_i) \left(\partial_j \phi^\star A_k - \partial_k \phi^\star A_j \right) \right. \\
&\quad + \tfrac{2}{3} \phi^\star A_i \, [A_j, A_k] \Big] .
\end{aligned}
\tag{13.22}
$$

These changes apply to the gravitational counterterm (13.18), namely

$$
I_g = \frac{1}{4\pi} \int_{M^n} \mathrm{Tr} \left(\phi^\star \omega \wedge d\phi^\star \omega + \frac{2}{3} \phi^\star \omega \wedge \phi^\star \omega \wedge \phi^\star \omega \right) .
\tag{13.23}
$$

Now we consider the equivalence relation $\pi (\phi) = \pi (\theta)$ for transformations strictly related to the mapping class group . We write π as

$$
\pi : \mathrm{Diff}^+ (M^n) \to \pi_0 \, \mathrm{Diff}^+ (M^n).
\tag{13.24}
$$

This allows us to verify that under large diffeomorphism transformations, the following relation holds:

$$
\mathcal{L}_{\mathrm{CSW}} \left(\phi^\star (A_i) \right) = I_g \left(\theta(\omega) \right).
\tag{13.25}
$$

Going back to the homeotopic term h_t, we consider a set of diffeomorphism transformations α, β, which obeys (13.24) and for which

$$
h_0 = 0; \quad h_1 = \beta.
$$

With these values specified, formula (13.25) takes the form

$$
\mathcal{L}_{\mathrm{CSW}} \left(\alpha (A) \right) = I_g \left(\beta(\omega) \right).
\tag{13.26}
$$

To show that equation (13.26) is invariant under $\pi_0 \, \mathrm{Diff}^+ (M^n)$, it suffice to show that it should be invariant under the variation

$$
\delta \mathcal{L}_{\mathrm{CSW}} \left(e^{i\phi}, e^{i\theta} \right) \simeq \delta I_g \left(e^{i(\phi + \epsilon)}, e^{i\theta} \right), \quad 0 < \epsilon < 2\pi.
\tag{13.27}
$$

Recall that h_t belongs to the (larger) homeotopy group, the mapping class group $\pi_0 \, \mathrm{Diff}^+ (M^n)$, so long as t is bounded by $0 \le t \le 1$. Consequently, h_t is none other than

$$
h_t \left(e^{i\phi}, e^{i\theta} \right) = \left(e^{i(\theta + t)}, e^{i\phi} \right),
\tag{13.28}
$$

and we easily see that mod 2π

$$\delta\mathcal{L}_{\mathrm{CSW}}\left((A_i)_{h_t}\right) \,=\, \delta I_g\left(\omega\right) \qquad\qquad (13.29)$$

the theory is mapping class group invariant. In other words there is no dis-connected general coordinate transformations that show up as we are varying the CSW effective action under large diffeomorphism transformations , and therefore, we conclude that Chern-Simons-Witten theories are global gravita-tional anomaly-free. In the process, we have recovered Witten's formula for the absence of global anomalies. (We have δ denoting large transformations in contrast to A in previous sections so as not to confuse it with the 1-form connection, A_i defining the CSW effective action.)

13.5 References

[1] Birman, J.: **Braids, Links, and Mapping Class Groups**, Princeton Annals of Mathematical Studies 82 Princeton University Press, Princeton, NJ 1975.

[2] Smale, S.: *Diffeomorphisms of the 2-Sphere*, Proc. Amer. Math. Society 10 (1959) 621-626.

[3] Cerf, J.: **Sur les Diffeomorphismes de la Sphère de Dimension Trois** ($\Gamma_4 = 0$), Lecture Notes in Mathematics Vol 53 Springer-Verlag 1968.

[4] Hatcher, A.: *A Proof of the Smale Conjecture*, Diff (S^3) $\simeq$ $O(4)$, Annals of Mathematics 117 (1983) 553-607.

[5] Kirby, R. and Siebenmann, L: *Foundational Essays on Topological Manifolds, Smoothings, and Triangulations*, Princeton Annals of Mathematical Studies 88 (1977).

- McCullough, D. and Miller, A.: **Homeomorphisms of 3-Manifolds With Compressible Boundary**, Memoirs American Mathematical Society, Vol. 61 Number 344 1986.

[6] Moise, E.: **Geometric Topology in Dimensions 2 and 3**, Graduate Texts in Mathematics, Vol. 47 Springer-Verlag, 1977.

[7] Alexander, J. W.: *An Example of a Simply Connected Surface Bounding a Region Which is Not Simply Connected*, Proc. Nat. Acad. Sciences

USA, Vol. 10 (1924) 6-8.

[8] Suzuki, S.: *On Homeomorphisms of 3-Dimensional Handlebody*, Canadian Journal of Mathematics Vol. 29 (1977) 111-124.

[9] Nielsen, J.: *Die Isomorphismmengruppe der Freien Gruppen*, Math. Ann. 91 (1924) 169-209.

[10] McCullough, D.: *Twist Groups of Compact 3-Manifolds*, Topology, Vol. 24 Number 4 (1985) 461-474.

[11] Waldhausen, F.: *Eine Klasse von 3-Dimensionale Manigfaltigkeiten* I, Invent. Math. 3 (1967) 303-333; and II (Idem) 4 (1967) 87-117.

- *On Irreducible 3-Manifolds Which Are Sufficiently Large*, Ann. Math. 87 (1968) 56-88.

[12] Kalliongis, J. and McCullough, D.: Pacific Journal Mathematics, Vol. 153 Number 1 (1992) 85-117.

[13] Swarup, G. A.: *Homeomorphisms of Compact 3-Manifolds*, Topology Vol. 16 (1977) 119-130.

[14] Baadhio, R. A.: *Global Gravitational Anomaly-Free Topological Field Theory*, Physics Letters B299 (1993) 37-40.

[15] Earle, C. and Eells, J.: Bulletin American Mathematical Society 73 (1967) 557.

[16] Baadhio, R. A.: *Knot Theory, Exotic Spheres and Global Gravitational Anomalies*, in **Quantum Topology**, Kauffman, L. H., and Baadhio, R. A. (Eds.). 78-90 (World Scientific) 1993.

[17] Baadhio, R. A. and Kauffman, L. H.: *Link Manifolds and Global Gravitational Anomalies*, Reviews in Mathematical Physics Vol. 5 Number 2 (1993) 331-343.

[18] Witten, E.: *Quantum Field Theory and the Jones Polynomial*, Communications in Mathematical Physics 121 (1989) 351-399.

[19] Witten, E.: *The Central Charge in Three Dimensions*, Phil. Trans. Royal Society London, A-Mathematical and Physical Sciences Series 329 No. 1605 (1989) 349-357.

[20] Baadhio, R. A.: *Mapping Class Groups for $D = 2+1$ Quantum Gravity and Topological Quantum Field Theories*, Nuclear Physics B441 (1995)

Nos. 1-2 383-401.

[21] Freed, D.: *Determinants, Torsions and Strings*, Communications Mathematical Physics 107 (1986) 483-513.

[22] Witten, E.: *Global Anomalies in String Theory*, in **Symposium on Anomalies, Geometry and Topology**, Eds. Bardeen, A. and White, A. p. 61-99 (World Scientific) 1985.

[23] Baadhio, R. A.: *Mapping Class Groups and Global Anomalies*, to appear.

[24] Witten, E.: *Global Gravitational Anomalies*, Communications Mathematical Physics 100 (1985) 197-229.

Chapter 14

Exotic Spheres

by Louis H. Kauffman

In Chapter 12, the role played by exotic spheres in the detection and cancellation of global anomalies was extensively analyzed. The purpose of this present chapter is to give a resumé (in the signature case) of the mathematical background involving characteristic classes that implies the existence of exotic spheres. To this end, we first review some basic facts about Chern classes, Pontrjagin classes, and the Hirzebruch index theorem. These facts are then marshalled to prove the existence of exotic spheres; in particular, the Milnor seven-sphere, Σ, and its relatives (see [1] for more information).

First, recall the infinite complex projective space $\mathbb{CP}^\infty$ and its interpretations for line bundles and cohomology: Let $[X, \mathbb{CP}^\infty]$ denote the homotopy classes of mappings of a space X to $\mathbb{CP}^\infty$. Then this homotopy set is isomorphic with the second cohomology group of X:

$$H^2(X) \cong \left[X, \mathbb{CP}_\infty\right].$$

This follows from the fact that $\mathbb{CP}_\infty$ is a $K(\mathbb{Z}, 2)$, a space whose homotopy groups all vanish except for a $\mathbb{Z}$ in dimension two.

It follows from the construction of $\mathbb{CP}^\infty$ that $\left[X, \mathbb{CP}^2\right] \cong \mathcal{L}(X)$, the isomorphism classes of complex line bundles over X. In this case, we have the canonical line bundle Λ over $\mathbb{CP}^\infty$, and a map $f : X \to \mathbb{CP}^\infty$ induces a line bundle $f^*\Lambda$ over X:

$$
\begin{array}{ccc}
f^\star \Lambda & \longrightarrow & \Lambda \\
\downarrow \pi & & \downarrow \pi^\star \\
X & \xrightarrow{f} & \mathbb{CP}^\infty.
\end{array}
$$

If $i \in H^2\left(\mathbb{CP}^\infty\right)$ denotes the generator of the cohomology ring of $\mathbb{CP}^\infty$, then the first Chern class of $f^\star\Lambda$, $c_1\left(f^\star\Lambda\right)$, is found by taking the pull-back of i via f:

$$
c_1\left(f^\star\Lambda\right) = f^\star(i) \in H^2\left(X\right).
$$

It is also not hard to see that $\mathcal{L}\left(X\right) = H^2\left(X\right)$ as groups, with tensor product of line bundles corresponding to addition in H^2. The first Chern class, c_1, can be interpreted as the self-intersection number of the 0-section of the corresponding bundle.

More generally, let $E \xrightarrow{p} B$ be a complex vector bundle. Then, there exist Chern classes $c_i\left(E\right) \in H^{2i}\left(B; \mathbb{Z}\right)$ satisfying the following properties:

14.0.1 Properties of Chern Classes

(0) $c_i\left(E\right) = 0$ for $i > n = $ complex fiber dimension of E.

$$
c(E) = 1 + c_1\left(E\right) + c_2\left(E\right) + \cdots + c_n\left(E\right),
$$

defines the total Chern class.

(1) If E and $\bar{E}$ are complex bundles isomorphic over B, then

$$
c(E) = c(\bar{E}).
$$

If $E \xrightarrow{p} B$ and $f : \bar{B} \longrightarrow B$, then $f^\star c\left(E\right) = c(f^\star \bar{E})$.

(2) $c\left(E \oplus \bar{E}\right) = c(E)\,c(\bar{E})$, where the product denotes cup product in the cohomology ring of B, and E and $\bar{E}$ are complex bundles over B.

(3) Let $\Lambda \longrightarrow \mathbb{CP}^\infty$ be the canonical line bundle. Then

$$
c_1\left(\Lambda\right) = i \in H^2\left(\mathbb{CP}^\infty\right)
$$

as described above. Similarly, if $\lambda \longrightarrow S^2$ is the canonical line bundle over S^2, then $c_1\left(\lambda\right) = g \in H^2\left(S^2\right)$ is the generator.

It is known (the splitting principle) that given a complex bundle $E \xrightarrow{p}$ B, then there exists a mapping $f : \bar{B} \longrightarrow B$ such that f^* injects the cohomology of B into the cohomology of $\bar{B}$ and $f^* E$ is a direct sum of line bundles. Thus we can write

$$f^* E \cong L_1 \oplus L_2 \oplus \cdots \oplus L_n$$

whence

$$\begin{aligned} f^* c(E) = c(f^* E) &= c(L_1) c(L_2) \cdots c(L_n) \\ &= \prod_{k=1}^{n} \left(1 + c_1 (L_k)\right). \end{aligned}$$

In this way, we see that the higher Chern classes can be expressed in terms of elementary symmetric functions of line bundles.

• **Example**

Let $E = \tau\, \mathbb{CP}_n$ = the tangent bundle to $\mathbb{CP}_n$. Explicitly,

$$\mathbb{CP}_n = S^{2n+1}/S^1$$

where S^1 is the unit complex numbers,

$$S^{2n+1} = \left\{ (z_0, z_1, \cdots, z_n) \in \mathbb{C}^{n+1} \mid |z_0|^2 + \cdots |z_n|^2 = 1 \right\}.$$

If $z = (z_0, z_1, \cdots, z_n)$ and $\lambda \in S^1$ then $\lambda z = \left(\lambda z_0\, \lambda z_1, \cdots, \lambda z_n\right)$. $E = \left\{ [u,v] \mid \|u\| = 1,\ u \cdot v = 0,\ (u,v) \sim \left(\lambda u, \lambda v\right) \right\}$. Here, $u, v \in S^{2n+1}$ and $[u,v]$ denotes the equivalence class of the pair (u,v) under the S^1-action.

Let $\Lambda_n \longrightarrow \mathbb{CP}_n$ denote the standard line bundle. Then,

$$\Lambda_n = \left\{ [u,\rho] \mid u \in S^{2n+1},\ \rho \in \mathbb{C},\ (u,\rho) \sim (\lambda u, \lambda \rho) \right\}.$$

Let $E' = \Lambda_n \oplus \Lambda_n \oplus \cdots \oplus \Lambda_n\ (n+1 \text{ copies})$. Then,

$$E' = \left\{ [u,v] \mid (u,v) \in S^{2n+1} \times \mathbb{C}^{n+1},\ (u,v) \sim (\lambda u, \lambda v) \right\}.$$

Hence, $E' \supset E$, and note that E' has the cross section $u \mapsto (u,v)$. Therefore, $E' \cong \tau\, \mathbb{CP}_n \oplus \epsilon$, where $\epsilon \to \mathbb{CP}^n$ denotes the trivial bundle in one complex dimension. We conclude that

$$c\left(\tau\, \mathbb{CP}^n\right) = c(E') = \left(1 + c_1 (\Lambda_n)\right)^{n+1}.$$

Letting $\alpha_n = c_1 (\Lambda_n)$ be the generator of $H^2 (\mathbb{CP}_n)$, we have the formula

$$c\left(\tau\, (\mathbb{CP}^n)\right) = (1 + \alpha_n)^{n+1}.$$

14.0.2 Pontrjagin Classes

If $E \xrightarrow{p} B$ is a real vector bundle, then we get an associated complex vector bundle $\hat{E} = E \otimes_{\mathbb{R}} \mathbb{C}$. Note that $\hat{E}$ and its complex conjugate bundle are isomorphic, i.e. $\hat{E}^{\star} \cong \hat{E}$. This implies that $2\, c_{2i+1}\left(\hat{E}\right) = 0$. We define the ith Pontrjagin class, $P_i\left(E\right)$, by the formula:

$$P_i\left(E\right) = (-1)^i\, c_{2i}\left(E \otimes_{\mathbb{R}} \mathbb{C}\right) \in H^{4i}\left(B\right),$$

and the total Pontrjagin class by the formula

$$P\left(E\right) = 1 + P_1\left(E\right) + \cdots + P_{[n/2]}\left(E\right),$$

where $[M]$ denotes the greatest integer in M. It then follows that

$$2\left(P\left(E \oplus E'\right) - P(E)\, P(E')\right) = 0.$$

The following Lemma (whose proof we omit) is useful.

Lemma 14.1 *1.) Let ω be a complex vector bundle. Then $\omega_{\mathbb{R}} \otimes \mathbb{C} \cong \omega \oplus \omega^{\star}$. (Here $\omega_{\mathbb{R}}$ denotes ω regarded as a real vector bundle.)*

2.) If $P_k = P_k\left(\omega_{\mathbb{R}}\right)$, $c_k = c_k\left(\omega\right)$, then $1 - P_1 + P_2 - \cdots \pm P_n = (1 - c_1 + c_2 - c_3 + \cdots \pm c_n)\left(1 + c_1 + \cdots + c_n\right)$.

- Example.

$\tau = \tau\, \mathbb{CP}^n$, $c(\tau) = (1 + a)^{n+1}$. $P_k = P_k\left(\tau_{\mathbb{R}}\right)$. Then $1 - P_1 + P_2 - \cdots = (1 - a)^{n+1}\,(1 + a)^{n+1} = (1 - a^2)^{n+1}$. Hence $1 + P_1 + \cdots + P_n = (1 + a^2)^{n+1}$. Hence $P_k\left(\mathbb{CP}^n\right) = \binom{n+1}{k}\, a^{2k}$.

Now we apply the Pontrjagin classes to study manifolds. Let M^{4n} denote a smooth, compact $4n$ manifold without boundary. Let $M^{4n} \in H_{4n}\left(M; \mathbb{Z}\right)$ denote the fundamental class of M^{4n}, and suppose that $i_1 + \cdots + i_r = n$, where $0 \le i_k \le n$. Let I denote the sequence $i_1, \cdots, i_r$ and define the Pontrjagin number $P_I\left[M^{4n}\right]$ by the formula

$$P_I\left[M^{4n}\right] = \left\langle P_{i_1} \cdots P_{i_r}, M^{4n}\right\rangle$$

where the brackets denote the evaluation of the product $P_{i_1}, \cdots P_{i_r}$ on the fundamental class. For instance, from our last example, we see that

$$P_I \left[\mathbb{CP}^{2n} \right] = \binom{2n+1}{i_1} \cdots \binom{2n+1}{i_r}.$$

The following theorem is basic to the relationship of Pontrjagin classes and cobordism.

Theorem 14.1 *If the smooth manifold M^{4n} is the boundary of a smooth $(4n+1)$-manifold B^{4n+1}, $M^{4n} = \partial B^{4n+1}$ then all Pontrjagin numbers $P_I\left(M^{4n}\right)$ vanish.*

Proof. Let μ_B denote the fundamental class in $H_{4n+1}\left(B, M\right)$. Then $\partial \mu_B = \mu_M$, where $\partial : H_{4n+1}\left(B, M\right) \to H_{4n}\left(M\right)$ is the homology boundary mapping. Furthermore, if $v \in H^{4n}\left(M\right)$ then $\langle v, \partial \mu_B \rangle = \langle \delta v, \mu_B \rangle$, where $\delta : H^{4n}(M) \to H^{4n+1}\left(B, M\right)$ is the coboundary map on cohomology. Now, we know that $\tau_B | M = \tau_M \oplus \epsilon$, hence $P_i\left(\tau_B | M\right) = P_i\left(\tau_M\right)$. It then follows directly from the exact sequence $H^{4n}\left(B\right) \xrightarrow{i^*} H^{4n}(M) \xrightarrow{\delta} H^{4n+1}\left(B, M\right)$ that $\delta\left(P_I\right) = 0$. Therefore,

$$\begin{aligned}
P_I\left(M^{4n}\right) &= \langle P_I, \mu_M \rangle \\
&= \langle P_I, \partial \mu_B \rangle \\
&= \langle \delta(P_I), \mu_B \rangle \\
&= 0.
\end{aligned}$$

This completes the proof ■

Thus we have shown that the $\mathbb{CP}^{2n}$ are not oriented boundaries. In fact, more is true. We can let Ω_n denote the oriented cobordism group of an n-dimensional smooth manifold. (Two oriented manifolds A^n and B^n are said to be cobordant if there exists an oriented $(n+1)$-manifold C^{n+1} such that $\partial C^{n+1} = A^n \cup \left(-B^n\right)$, where $-B^n$ denotes B^n with the reverse orientation. A manifold, A^n, is cobordant to $\varnothing$ if B^n can be taken to be empty. $\varnothing$ produces an inverse in cobordism classes since $\delta\left(A^n \times I\right) = A^n \cup \left(-A^n\right)$ and it is easy to see that the connected sum $A^n \cup B^n$ is cobordant to the connected sum $A^n \sharp B^n$. Thus, $A^n \sharp \left(-A^n\right)$ is cobordant to $\varnothing$.)

Ω_n is a ring with addition the operation of connected sum ($\sharp$) and multiplication the cartesian product. It is known that Ω_n is finite for $n \not\equiv 0$ (mod 4) and that $\Omega_{4k} \otimes \mathbb{Q}$ has a basis

$$\left\{ \mathbb{CP}^{2i_1} \times \cdots \times \mathbb{CP}^{2i_r} \mid I = i_1 i_2 \cdots i_r \text{ is a partition of } 4k \right\}.$$

See [M] for a proof of this result (due originally to René Thom).

We are now in a position to state and prove the fundamental theorem of Hirzebruch, connecting the signature of a $4k$-manifold with its Pontrjagin classes. The idea is to produce combinations of Pontrjagin classes that behave formally like the signature, and then use cobordism theory to check agreement on the relevant examples.

Recall the important properties of the signature, $\sigma \left(M^{4k} \right)$:

1) By definition, $\sigma \left(M^{4k} \right)$ is the signature of the quadratic form

$$\begin{aligned} H^{2k}(M) \times H^{2k}(M) &\longrightarrow \mathbb{Z} \\ a, b &\longmapsto \langle a \cup b, [M] \rangle. \end{aligned}$$

2) $\sigma \left(M_1 + M_2 \right) = \sigma \left(M_1 \right) + \sigma \left(M_2 \right)$ where $M_1 + M_2 = M_1 \sharp M_2$, the connected sum.

3) If $M^{4k} = \partial N^{4k+1}$ then $\sigma \left(M^{4k} \right) = 0$.

Thus if M_1^{4k} is cobordant to M_2^{4k}, then $\sigma \left(M_1^{4k} \right) = \sigma \left(M_2^{4k} \right)$.

4) $\sigma \left(M_1^{4k} \times M_2^{4k} \right) = \sigma \left(M_1^{4k} \right) \sigma \left(M_2^{4k} \right)$.

Thus $\sigma : \Omega_* \longrightarrow \mathbb{Z}$ is a homomorphism from the cobordism ring to the integers. The Pontrjagin numbers already obey 2) and 3). We need to cook up property 4). For this, we need the concept of a multiplicative sequence: Let R be a commutative ring with unit, 1. Let $A^* = (A^0, A^1, A^2, \cdots)$ be a graded R-algebra. Let $A^\pi = \left\{ a_0 + a_1 + a_2 + \cdots \mid a_i \in A^i \right\}$ be the associated formal power series ring. Let $K_i \left(x_1, x_2, \cdots, x_i \right)$ be a sequence of polynomials such that each K_n is homogeneous of degree n. Let

$$K : A^\pi \longrightarrow A^\pi \text{ via } K(a) = 1 + K_1(a_1) + K_2(a_1, a_2) + K_3(a_1, a_2, a_3) + \cdots.$$

We say that K is multiplicative if $K(ab) = K(a)K(b)$ for all $a, b \in A^\pi$.

Lemma 14.1 *Given a formal power series* $f(t) = 1 + \lambda_1 t + \lambda_2 t^2 + \cdots$, *there exists a unique multiplicative sequence* $\{K_n\}$ *such that* $K(1+t) = f(t)$.

Proof.

For uniqueness, let $A^\star = R[t_1, t_2, \cdots, t_n]$ and $\sigma = (1 + t_1)(1 + t_2) \cdots$ with $\sigma_1, \sigma_2, \cdots, \sigma_n$ the elementary symmetric functions so that

$$\sigma = 1 + \sigma_1 + \sigma_2 + \cdots + \sigma_n.$$

Then

$$K(\sigma) = K(1+t_1) K(1+t_2) \cdots K(1+t_n) = f(t_1) f(t_2) \cdots f(t_n).$$

Thus $K(\sigma_1, \sigma_2, \cdots, \sigma_n)$ is uniquely determined by $f(t)$. Since $\sigma_1, \sigma_2, \cdots, \sigma_n$ are algebraically independent, this proves uniqueness.

For existence, let $I = i_1 i_2 \cdots i_r$ be a partition of k and define

$$S_I(\sigma_1, \cdots, \sigma_n) = \sum t_1^{i_1} t_2^{i_2} \cdots t_r^{i_r}$$

where this sum means that we sum over all choices of r-subsets, thereby obtaining a symmetric function and hence a polynomial in the elementary symmetric functions $\sigma_1, \sigma_2, \cdots, \sigma_n$. These polynomials form a basis for the symmetric homogeneous polynomials of degree k in the variables $t_1, t_2, \cdots, t_n$. Thus, letting $\lambda_I = \lambda_{i_1} \lambda_{i_2} \cdots \lambda_{i_r}$, we can write

$$K_n(\sigma_1, \cdots, \cdots \sigma_n) = \sum_I \lambda_I S_I(\sigma_1, \cdots, \sigma_n)$$

where I ranges over all partitions of n. It follows that

$$S_I(ab) = \sum_{HJ=I} S_H(a) S_J(b),$$

where HJ denotes the partition obtained by juxtaposition. Hence

$$\begin{aligned} K(ab) &= \sum_I \lambda_I S_I(ab) \\ &= \sum_I \lambda_I \sum_{HJ=I} S_H(a) S_J(b) \\ &= \sum_{H,J} \lambda_H S_H(a) \lambda_J S_J(b) \\ &= K(a) K(b). \end{aligned}$$

This completes the proof ∎

Now let $\{K_n (x_1 \cdots x_n)\}$ be a multiplicative sequence of polynomials with rational coefficients. Let M^{4k} be a smooth compact oriented $4k$-manifold. Define the K-genus of M^{4k} by the formula

$$K[M^{4k}] = K_k[M^{4k}] \langle K_k (P_1, \cdots, P_k), [M^{4k}]\rangle,$$

where P_i denotes the ith Pontrjagin class of τ_M. If $4 \nmid$ does not divide dim (M), define $K[M] = 0$.

Lemma 14.1 *If $\{K_n\}$ is any multiplicative sequence with rational coefficients, then the correspondence $M \mapsto K[M]$ defines a ring homomorphism $\Omega_* \longrightarrow \mathbb{Q}$ and hence an algebra homomorphism*

$$\Omega_* \otimes \mathbb{Q} \longrightarrow \mathbb{Q}.$$

Proof. We need only check the behavior on products. $M \times M'$ has total Pontrjagin class $P \times P'$ modulo elements of order 2. So $K((P \times P')) = K(P) \times K(P')$ and

$$\langle K(P) \times K(P'), \mu \times \mu'\rangle = (-1)^{mm'} \langle K(P), \mu\rangle \langle K(P'), \mu'\rangle.$$

Hence, $K[M \times M'] = K[M]\, K[M']$ ∎

Now we can state and prove the Hirzebruch index theorem.

Theorem 14.1 (Hirzebruch) *Let $\{L_k\}$ be the multiplicative sequence of polynomials corresponding to $f(t) = \sqrt{t}/\tanh\left(\sqrt{t}\right)$. Then*

$$\sigma\left(M^{4k}\right) = L[M^{4k}].$$

Proof. By the quoted result on $\Omega_{4k} \otimes Q$, it suffices to check the theorem for $L_k[\mathbb{CP}^{2k}]$. Here $P = (1 + a^2)^{2k+1}$. Since $L(1 + a^2) = \sqrt{a^2}/\tanh\left(\sqrt{a^2}\right)$, $L(P) = (a/\tanh a)^{2k+1}$. Hence $L[\mathbb{CP}^{2k}] = \langle L(P), \mu\rangle$ equals the coefficient of a^{2k} in $(L(1 + a^2))^{2k+1}$. We check this coefficient by residues. Let

$$u \tanh(z) = \left(e^z - e^{-z}\right) / \left(e^z + e^{-z}\right).$$

Then $du = (1 - u^2)dz$ whence

$$dz = \frac{du}{1 - u^2} = (1 + u^2 + u^4 + \cdots)\, du.$$

$$\therefore\ L[\mathbb{CP}^{2k}] = \frac{1}{2\pi i} \oint \frac{dz}{z^{2k+1}} \left(\frac{z}{\tanh z}\right)^{2k+1}$$

$$= \frac{1}{2\pi i} \oint \left(\frac{dz}{\tanh z}\right)^{2k+1}$$

$$= \frac{1}{2\pi i} \oint \frac{(1 + u^2 + u^4 + \cdots)}{u^{2k+1}}\, du$$

$$= 1.$$

Hence, $L[\mathbb{CP}^{2k+1}] = 1 = \sigma\left(\mathbb{CP}^{2k}\right)$. This completes the proof ■

Here are some useful facts about the series $\sqrt{t}/\tanh\left(\sqrt{t}\right)$:

$$\sqrt{t}/\tanh\left(\sqrt{t}\right) = 1 + \frac{1}{3}t - \frac{1}{45}t^2 + - \cdots + \left(-1^{k-1}\, 2^{2k}\, \frac{B_k\, t^k}{(2k!)!} + \cdots\right)$$

where B_k is the kth Bernoulli number. The first few L-polynomials are:

$$L_1 = \tfrac{1}{3}P_1,$$
$$L_2 = \tfrac{1}{45}\left(7P_2 - P_1^2\right),$$
$$L_3 = \tfrac{1}{945}\left(62P_3 - 13P_1 P_2 + 2\,P_1^3\right).$$

14.1 Exotic Spheres

The example that we are about to discuss is not the first example of an exotic differentiable structure on a sphere, but it is diffeomorphic to that example. The first example is due to Milnor [2] and produces a non-standard differentiable structure on a sphere of dimension seven. The example we are about to discuss is due to Brieskorn [3].

The Brieskorn examples arise from studying algebraic varieties associated with polynomials of the form

$$f(z) = z_0^{a_0} + z_1^{a_1} + \cdots + z_n^{a_n}$$

where $a_0, a_1, \cdots, a_n$ are positive integers and the z_i's are complex variables.

Let $V(f)$ denote the variety of f:

$$V(f) = \{z \in \mathbb{C}^{n+1} | f(z) = 0\}.$$

Let $K(f)$ denote the intersection of this variety with the unit sphere in $\mathbb{C}^{n+1}$:

$$\begin{aligned} K(f) &= V(f) \cap S^{2n+1} \\ &= \{z \in \mathbb{C}^{n+1} | f(z) = 0 \text{ and } |z| = |z_0|^2 + \cdots + |z_n|^2 = 1\}. \end{aligned}$$

It is not hard to check that $V(f)$ is a manifold away from $\vec{0} \in V(f)$ and that the intersection of $V(f)$ with S^{2n+1} is transversal. Hence, $K^{2n-1}(f)$ is a smooth manifold of dimension $2n - 1$.

Under these conditions, the manifolds $K^{2n-1}(f)$ are sometimes homeomorphic to spheres and sometimes cannot be diffeomorphic to standard spheres. A case in point is $K^7(f)$ for $f = z_0^3 + z_1^5 + z_2^2 + z_3^2 + z_4^2$. In general, let $\Sigma(a_0, a_1, \cdots, a_n)$ denote $K(f)$ for $f = z_0^{a_0} + z_1^{a_1} + \cdots + z_n^{a_n}$. Thus we assert that $\Sigma(3, 5, 2, 2, 2) = \Sigma^7$ is an exotic sphere. We shall finish this chapter with a number of different points of view on this fact. Here are the facts that we will show:

1) $\Sigma^7 = \Sigma(3, 5, 2, 2, 2)$ is the boundary of a smooth 8-manifold of signature -8: $\Sigma^7 = \partial N^8$, $\sigma(N^8) = -8$.

2) Σ^7 is homeomorphic to a 7-dimensional sphere.

With these facts in hand, the exoticity of Σ^7 is proved as follows: An extra fact about the manifold N^8 is that it is connected and has vanishing homology except in dimension four. We can form the topological manifold $M^8 = N^8 \cup_\Sigma D^8$ where D^8, is a standard 8-ball. If M^8 is a smooth manifold, then $\sigma(M^8) = L[M^8]$. But $\widetilde{H}^*(M) = 0$ for $\star \neq 4, 8$.

$$\begin{aligned} P_1 &\in H^4(M^8) \\ P_2 &\in H^8(M^8); \end{aligned}$$

and we have

$$\sigma(M^8) = L_2\left(M^8\right) = \frac{7}{45}\left[P_2(M^8) - P_1^2(M^8)\right];$$

$$-8 = \frac{7}{45}\left[P_2(M^8) - P_1^2(M^8)\right];$$

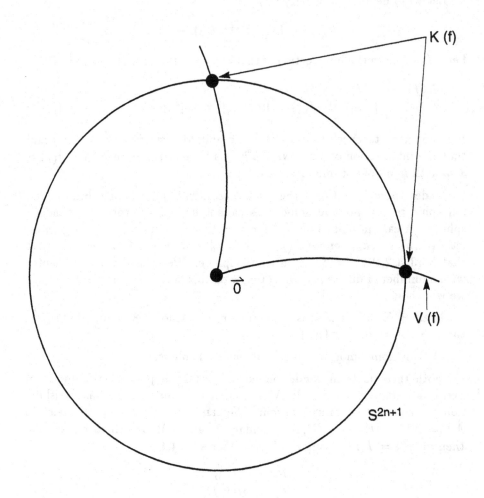

Figure 14.1:

$$-8 \cdot 45 = 7 \left[P_2(M^8) - P_1^2(M^8) \right];$$

$$-2^3 \cdot 3^2 \cdot 5 = 7 \left[P_2(M^8) - P_1^2(M^8) \right].$$

Since $(P_2(M^8) - P_1^2(M^8))$ is an integer and 7 does not divide $-2^3 \cdot 3^2 \cdot 5$, we conclude that M^8 does not have a differentiable structure. Since Σ diffeomorphic to S^7 would allow a differentiable structure on M^8, this shows that Σ is not diffeomorphic to S^7. Thus Σ is an exotic sphere.

For the record, Milnor's original example [2] of an exotic 7-sphere was constructed as follows: For each $(h, j) \in \mathbf{Z} \oplus \mathbf{Z}$ let $f_{h_j} : S^3 \longrightarrow \mathrm{SO}(4)$ be defined by the equation $f_{h_j}(u) \cdot v = u^h v u^j$ for $v \in \mathbb{R}^4$. Here we take quaternion multiplication on the right. Let ξ_{h_j} denote the 3-sphere bundle over S^4 determined by the map f_{h_j}. That is, with $S^4 = D_+^4 \cup_{S^3} D_-^4$, the quantity ξ_{h_j} is equivalent to $D_\pm^4 \times S^3$ over $D_\pm^4$ and f_{h_j} provides the pasting data for gluing these two trivial bundles to form ξ_{h_j}. Let M_k^7 denote the total space of the bundle ξ_{h_j} where $h + j = 1$ and $h - j = k$. Milnor shows that M_k^7 is homeomorphic to S^7 for all k and that M_k^7 is exotic when $k^2 \not\equiv 1 \pmod 7$. The argument involves the Pontrjagin classes of the bundle.

Now let us return to the Brieskorn manifolds and discuss some aspects of their structures. Consider $f(z) = z_0^{a_0} + z_1^{a_1} + \cdots + z_n^{a_n}$ as a mapping $f : \mathbb{C}^{n+1} \longrightarrow \mathbb{C}$. It is easy to see that $f | \mathbb{C}^{n+1} - V(f) : \mathbb{C}^{n+1} - V(f) \longrightarrow \mathbb{C} - \{0\}$ is a fiber bundle, and that by taking the restriction to $E_\delta = f^{-1}(S_\delta^1) \overset{f}{\longrightarrow} S_\delta^1$ where $S_\delta^1 = \{z \in \mathbb{C} | |z| = \delta\}$ for δ small, we also get a fiber bundle and that $E_\delta \cap D^{2n+2} \longrightarrow S_\delta^1$ gives a fiber bundle with the boundary of each fiber diffeomorphic to $K(f)$. Milnor [9] generalized this fiber bundle structure to a bundle $\phi : S^{2n+1} - K(f) \longrightarrow S^1$, $\phi(z) = f(z)/|f(z)|$. In the case of $E_\delta \cap D^{2n+2} \longrightarrow S_\delta^1$ and $\phi : S^{2n+1} - K(f) \longrightarrow S^1$ are equivalent bundles by using the mapping

$$(z_0, z_1, \cdots, z_n) \longrightarrow \left(\rho^{1/a_0} z_0, \rho^{1/a_1} z_1, \cdots, \rho^{1/a_n} z_n \right)$$

for ρ real (choosing ρ so that the image point is on the sphere). In the general case (of f with an isolated singularity at the origin) Milnor uses a vector field to push the fibers of $E_\delta \cap D^{2n+2}$ out into the sphere.

A similar bit of geometric topology lets us see that $K\left(x^k + f(z)\right)$ is a k-fold branched cyclic cover of S^{2n+1} branched along $K(f)$. This sets the stage separate sets of variables) in terms of $K(f) \subset S^{2n+1}$ and $K(g) \subset S^{2m+1}$, where $f = f(z_0, \cdots, z_n), g = g(z_0, \cdots, z_m)$. Here, the idea is as follows. Suppose we are given maps $f : D^{2n+2} \longrightarrow D^2$ and $g : D^{2m+2} \longrightarrow D^2$ with singular fiber bundles elsewhere. Then we can form the pull-back $Z = \{(x, y) \in D^{2n+2} \times D^{2m+2} | f(x) = g(y)\}$:

$$
\begin{array}{ccc}
Z & \longrightarrow & D^{2m+2} \\
\downarrow & & \downarrow g \\
D^{2n+2} & \overset{f}{\longrightarrow} & D^2.
\end{array}
$$

and $\partial Z \hookrightarrow \partial(D^{2m+2} \times D^{2n+2}) \cong S^{2(n+m)+3}$. Appropriate analysis shows that $\partial Z \subset S^{2(n+m)+3}$ is equivalent to $K(f+g) \subset S^{2(n+m)+3}$. See [4].

For example, in the case of $x^k + f(z)$, we have

$$
\begin{array}{ccc}
Z & \longrightarrow & D^2 \\
\downarrow & & \downarrow \\
D^{2n+2} & \overset{f}{\longrightarrow} & D^2.
\end{array}
$$

with $g(x) = x^2$. Here it is easy to see that ∂Z is the k-fold cyclic branched covering of S^{2n+1} along $K(f)$. Note that this construction gives a canonical embedding in a sphere of two dimensions higher.

Thus, if $K^{n-2} \subset S^n$ then we have $K^n_a \longrightarrow S^n$ as branched cover, and $K^n_a \subset S^{n+2}$ where K^n_a denotes the a-fold cyclic branched cover of S^n.

In this way, we get an inductive definition of the Brieskorn manifolds as iterated branched coverings. $\Sigma(a_0, a_1)$ is a torus link of type (a_0, a_1) in S^3. For example $\Sigma(3, 5) \subset S^3$ has diagram sketched in Figure 14.2.

Our Milnor sphere $\Sigma(3, 5, 2, 2, 2) \subset S^9$ is the result of three 2-fold branched coverings starting from the $(3, 5)$ torus knot.

$$
K_{3,5} \subset S^3 \longleftarrow K_{3,5,2} \subset S^5 \longleftarrow K_{3,5,2,2} \subset S^7.
$$

These constructions give a clear view of the algebraic topology of the bounding manifolds. We shall only sketch these details here, referring the reader to [3], [4], and [5].

Given $K(f) \subset S^{2n+1}$ we have that $K(f) = \partial N(f)$ is the fiber of the Milnor fibration alluded above. The intersection form of middle dimension of

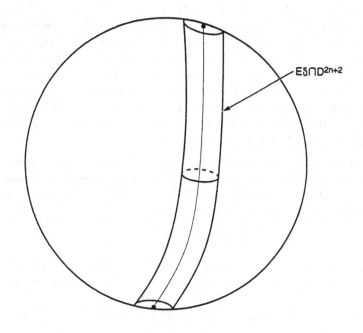

$E \cap D^{2n+2}$

Figure 14.2:

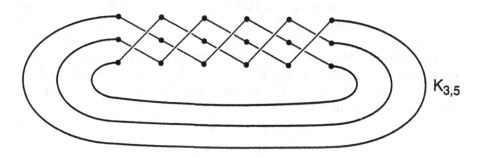

$K_{3,5}$

Figure 14.3:

Given $K(f) \subset S^{2n+1}$ we have that $K(f) = \partial N(f)$ is the fiber of the Milnor fibration alluded above. The intersection form of middle dimension of the homology of $N(f)$ is given by $\theta(f) \pm \theta(f)^{\mathsf{T}}$ where $\theta(f) : H_{n+1}(N(f)) \times H_{n+1}((f)) \longrightarrow \mathbb{Z}$ is the Seifert linking pairing obtained by the formula $\theta(f)(a, b) = lk(a^{\star}, b)$ where lk denotes the linking number and $a^{\star}$ is the cycle in $S^{2n+1} - N(f)$ obtained by pushing a along a positive normal to $N(f)$ into the complement. One finds that $\theta(f + g) \cong \theta(f) \otimes \theta(g)$ and consequently it is easy to determine intersection forms for composites. In particular, one has $\theta(f + x^2) \cong \theta(f)$.

The construction we have discussed generalizes to a *tensor product construction* for $K^n \subset S^{n+2}$, $L^m \subset S^{m+2}$ (L is a fibered codimension two submanifold of S^{m+2} to $(K \otimes L)^{n+m+1} \subset S^{n+m+3}$). Thus, we can start with any knot $K \subset S^3$ and form

$$[K \otimes \Sigma(2, 2, 2)]^7 \subset S^9.$$

If θ is a Seifert pairing for K in S^3, then $K \otimes \Sigma(2, 2, 2) = \partial \mathcal{N}$, $\mathcal{N} \subset S^9$ with the same Seifert pairing. As a result, $\mathcal{N}$ has intersection pairing $\theta + \theta^{\mathsf{T}}$ and hence

$$\sigma(\mathcal{N}) = \sigma\left(\theta + \theta^{\mathsf{T}}\right) = \sigma(K),$$

the classical signature of the knot. As a consequence, many exotic spheres can be constructed directly in relation to knots and links in S^3.

The manifolds $K \otimes \Sigma(2, 2, 2, \cdots, 2)$ (n 2's) admit actions of the orthogonal group $O(n)$ with orbit space D^4 and fixed point set $K \subset S^3 = \partial D^4$. These are called link manifolds and are classified in [6]. We have discussed their relationship with global anomalies in [7].

It is also worth pointing out that the Brieskorn manifolds are tensor products of *empty knots* $[a] : \Sigma(a_0, a_1, \cdots, a_n) = [a_0] \otimes [a_1] \otimes \cdots, \otimes [a_n]$ where $[a] : S^1 \longrightarrow S^1$, $[a](\lambda) = \lambda^a$. The term $[a] : S^1 \longrightarrow S^1$ is a fibration corresponding to the empty knot $\psi \subset S^1$ (the empty set has dimension -1). By looking at the inverse image of a point in S^1 under $[a]$, we get the fiber consisting of discrete points (a in number), and hence the Seifert pairing of this empty knot. It has the form:

$$\begin{pmatrix} 1 & 0 & 0 \\ -1 & 1 & 0 \\ 0 & -1 & 1 \end{pmatrix}$$

Finally, we should mention that so far we have only mentioned exotic spheres that are boundaries of parallelizable manifolds. There is a big class of exotic differentiable structures that do not bound in this way. Their properties require homotopy theory for detection. Such very exotic n-spheres are classified by $\pi_{n+k}(S^k)/\mathrm{Im}(J)$, where $\pi_{n+k}(S^k)$ denotes a stable homotopy group of the sphere S^k, and $\mathrm{Im}(J)$ denotes the image of the J- homomorphism:

$$J: \pi_n(\mathrm{SO}(k)) \longrightarrow \pi_{n+k}(S^k).$$

See [8] for more information on these matters.

Very exotic spheres may have some physical relevance, according to a conjecture by Witten. In reference [10] he postulates that gravitational instantons and/or solitons have the structure of very exotic spheres. Our knowledge of gravitational instantons and solitons is limited, but there is no doubt that deeper knowledge about very exotic spheres should shed light on these relationships. In addition to reference [10] where the role of very exotic spheres is detailed for the case of ten-dimensional supergravity theories, the interested reader may want to consult reference [11], [12], and [13] where the contributions to superstring theory of very exotic spheres are studied.

14.2   References

[1] Milnor, J. W. and Stasheff, J. D.: **Characteristic Classes**, Annals of Mathematical Studies No. 76, Princeton University Press 1974.

[2] Milnor, J. W.: *On Manifolds Homeomorphic to the 7-Sphere*, Annals of Mathematics 64 (1956) 399-405.

[3] Brieskorn, E.: *Beispiele zur differential topologic von singularitäten*, Invent. Math. 40 (1966) 153-160.

[4] Kauffman, L. H. and Neumann, W. D.: *Products of knots, branched fibrations and sums of singularities*, Topology 16 (1977) 369-393.

[5] Kauffman, L. H.: **On Knots**, Annals of Mathematical Studies No. 115, Princeton University Press 1987.

[6] Kauffman, L. H.: *Link Manifolds*, Michigan Mathematics Journal 21 (1974) 33-44.

[7] Baadhio, R. A. and Kauffman, L. H.: *Link Manifolds and Global Gravitational Anomalies*, Reviews in Mathematical Physics Vol. 5 No. 2 (1993) 331-343.

[8] Kervaire, M. A. and Milnor, J. W.: *Groups of Homotopy Spheres I*, Annals of Mathematics, Vol. 77 No. 3 (1963) 504-537.

[9] Milnor, J. W.: **Singular Points of Complex Hypersurfaces**, Annals of Math. Study 6 Princeton University Press 1968.

[10] Baadhio, R. A. and Lee, P.: *On the Global Gravitational Instanton and Soliton that are Homotopy Spheres*, Journal of Mathematical Physics Vol.32 No. 10 (1991) 2869-2874.

[11] Baadhio, R. A.: *Global Gravitational Instantons and their Degrees of Symmetry*, Journal of Mathematical Physics Vol. 33 No. 2 (1991) 721-724.

[12] Baadhio, R. A.: *On Global Gravitational Instantons in Superstring Theory*, Journal of Mathematical Physics, Vol. 34 No. 2 (1993) 358-368.

[13] Baadhio, R. A.: *Vacuum Configuration for Inflationary Superstring*, Journal Mathematical Physics 34 No. 2 (1993), 345-357.

Index